The Environment of Change

The Environment
of Change

AARON W. WARNER

DEAN MORSE *and*

THOMAS E. COONEY

Editors

Columbia University Press

NEW YORK AND LONDON   1969

Copyright © 1969 Columbia University Press
Library of Congress Catalog Card Number: 79-79572
Printed in the United States of America

TO FRANK TANNENBAUM, *Director of the
University Seminars of Columbia University,
whose dedication to the task of building
paths of communication between disciplines
and institutions was the force responsible
for this volume and the conferences of
which it is a reminder.*

PREFACE

THE PAPERS AND DISCUSSIONS presented in this book represent the proceedings of two conferences sponsored by *Time* magazine under my predecessor as publisher, Bernhard M. Auer. Their purpose was to bring together members of the academic and business communities who shared a concern about the impact of science and technology on society.

The meetings were organized by the steering committee of the Columbia University Seminar on Technology and Social Change, a group which holds regular monthly meetings of university faculty members, business leaders, and public officials. Drawing upon its impressive roster of members the Seminar arranged to have some of its number deliver papers or to otherwise participate in the *Time* conferences, the first held at the Onchiota Conference Center, Sterling Forest, New York, in January of 1966, and the second at Airlie House, Warrenton, Virginia, the following September.

The speakers were asked to address themselves at will to the general subject of social change. Sir Isaiah Berlin, noted for his analyses of Marxism and Russian Communism, chose to talk about the dangers of putting millenarian theories of revolution into practice. Professor I. I. Rabi, a Nobel Laureate physicist, focused on the need for an understanding of the scientific roots of technological change. The Salk Institute's Jacob Bronowski, reviewed the accomplishments of biological technology—from organ transplants to the control of heredity. Economist Eli Ginzberg considered the changes wrought on employment patterns, on personal and social values, and on behavior, by modern technology. Professor Everett Kassalow reviewed the role of technology in developing nations and its use by industrial nations in assisting them. Professor Loren Eiseley, as an anthropologist, used cultural comparisons to call into question the assumption of modern industrial man that the good life is only obtainable by building a technological society.

The formal papers were followed by questions from the floor and informal discussion between speaker and audience. These discussions continued into meal times and frequently well past them. The use in this volume of colloquies not directly identified with the responses to a particular speaker's remarks reflects these longer, and more informal, sessions.

July, 1968

James R. Shepley
Publisher, *Time*

INTRODUCTION

THE EDITORS OF THIS VOLUME have been faced with a task which was perhaps impossible. The more they edited the hundreds of pages of argument and counter-argument, dry comments and, on occasion, impassioned pleas of two conferences devoted to "The Environment of Change," the more they ran the danger that the very act of excising material, clarifying ideas, and ordering argument would destroy what they were most intent upon preserving—the lively sense of search, the frank admission of uncertainty, the fullness of dialogue. The drama of the two events, so clear to the participants, could not be rendered properly without the gifts of a Plato.

Fundamentally, then, what we, the editors, should have been able to produce was a kind of contemporary *Symposium* in which many voices would clearly be heard. But to accomplish this task properly we would have had to listen to many different groups talking simultaneously throughout the two conferences, noting the dialogues which began at breakfast, listening to voices long after midnight, able indeed to hear the silent voices of those who found that the best place to discover their thoughts was along the borders of a lake in the foothills of the Catskills, where the first conference was held, or in contemplation of the meadowland of Virginia, where some of us came together again eight months later, joined by a new group.

Some of us were called "faculty" and some called themselves "students," but of course neither conference was really a conference of conventional faculty or students. Those who might most aptly have been called "faculty"—the half dozen who were invited to give talks—were not to be found at the usual college or university. This "faculty" had experienced direct involvement in the affairs of what Henry James called the "great world," the arena of action and will, serving for a time the state, corporations, or the great institutions of labor. Another group of "faculty" were several members of the Columbia University Seminar on Technology and Social Change, who had met together once a month for several years to try to map the shifting boundaries of the area of impact of modern technology, modern science, and those new institutions which have been created to direct and control scientific activity and technological change.

Finally, the "students"—who were they? Some were businessmen from varied groups of industries and with varied roles within those industries; the rest were members of the staff of *Time* Magazine. "Faculty" and "students," we were some forty people and we were brought together for three days at each conference. The logistics of the conferences were handled in a skillful and unobtrusive way by the public relations staff of *Time* Magazine. Their intention, happily executed, was to provide the group with an environment of calm, a place where the predominant motion was that of minds reflecting on the swirl of change.

Having provided the means for the conferences, the magazine's publisher turned over to the steering committee of the Columbia University Seminar on Technology and Social Change the task of selecting subjects which might properly be considered under the general theme, and people who would be invited to talk informally to the conference. The authors of the papers in this volume are those we selected for this purpose. They were chosen because we believed that their breadth of view, the range of their activities, the importance of their contributions to scientific, political, historical, esthetic, and social understanding could not but provide the conference with stimulation and guidance. In a sense they were our "authorities," but quickly their authority became unobtrusive and natural! The arrangements of the conference saw to that. No one could tell whether the accidents of early arrival at the breakfast table would place him next to a Nobel prize winner in physics or a poet-mathematician at the Salk Institute or an engineer-architect-mathematician from Columbia University. And over coffee in the morning or sharing a beer late at night, community of concern and interest was found. If there is one quality of the conference that stood out, it is that somehow we were not indifferent to one another. We were brought together by common interests and concerns.

At each conference there were five papers given by these invited guests, some of them quite informal in character. Following the talks, the entire group discussed the subject at some length, many of these discussions being resumed in the evening and going on to all hours. They were far-ranging, often impassioned, in character. Neither authority nor age nor position guaranteed immunity from criticism. A businessman might tell a physicist that business, since it provided much of the funds for scientific research, should have a dominant say in how these funds should be used, implying in no uncertain terms that he felt that business would be more responsive to social needs and less wasteful of resources, only to be told, no less emphatically, that business in general was hopelessly ill-equipped by knowledge and training for such a supervisory role. The frequent plea from the citizen that science and technology should serve the social interest, and that only the citizen himself could say what that interest was, met the equally passionate response that scientific advance has its own logic and justification, that a science subservient to these very social needs would be no science at all in a very short time. The flavor of such raw confrontations, sudden, sharp, stimulating, even if the problem were to remain unresolved, provided much of the excitement of the conferences. Differences of opinion were the order of the day, and such differences were not confined to the opposing views of academia and the "great world." Just as often conflicts arose between physical scientist and social scientist, perhaps about the very nature of scientific activity itself. Again it might be two businessmen expressing apparently irreconcilable views about the proper role of the corporation in a society of rapid technological change, or perhaps a disagreement of a radical sort about the pace of technological advance. Agreement was not the usual state of affairs.

But perhaps an agreement that did exist was really more important. Busy men, hard pressed to steal three days' time from the concerns of large-scale business, scholars whose usual activities were teaching and research, agreed to listen to each other, agreed that it was important that they sit down together, agreed that time had to be found, not to bridge the two cultures of a C. P. Snow, but rather to bridge a different kind of gap. The gap that Snow has called our attention to has arisen in spite of quite close physical propinquity. The physicist and the humanist of the university do, we should not forget, in all probability live in the same community. If they fail to communicate, it is often because their concerns are different, their special languages inaccessible to anyone outside their own discipline. They avoid talking to each other because bitter experience has, all too frequently, taught them that they talk past one another. They are afraid of being misunderstood or, even worse, of not understanding. The gap between the "great world" and academia seems, at least in part, to be a different kind of phenomena. It seems to be in part simply a geographical fact. Academia is here; the world of business and politics is there. We do not meet often, and when we do meet it is apt to be at ceremonial occasions or at times when very limited ends bring us together.

We are not often brought together where the purpose of the convening is just that, to be together. It was *Time* Magazine's gift to us that as far as possible we were brought together just because it seemed a good idea for us to live, eat, reflect, and discuss with each other.

We were not trying to solve problems. The title of the conference itself did not state a problem. Problems, it is true, might be introduced, but solutions were not expected. We wanted, in short, to see what kinds of problems did trouble us, individually and collectively. The only general assertion was that an environment of change does have implications, cannot fail to affect us in profound ways, morally and intellectually, socially and politically. But an environment of change can become too easily accepted. It can become the water we swim in so, like the fish, we cease to be aware of it. But environmental change is not costless, and we do strive to reduce it to the predictable and orderly; we assume that we control it, that we plan it, that we are its masters, and that we have chosen it.

Isaiah Berlin's introductory remarks, at the first conference and in the discussion that followed, seemed to set the tone for much of our subsequent discussion. He dealt with the great historical problems of prediction and planning, the brave but pathetic efforts of some of the most devoted students of human affairs to reduce to a pattern the environment of change of the last few centuries—after all the present has no monopoly on rapid technological and social change. If change can be reduced to the explicable and the orderly, then much of its capacity to inspire anxiety, even terror, is disarmed. But Sir Isaiah cautioned us against easy prophetic views. An obsession with the distant future, an attempt to plan too thoroughly for this future, a belief that we know exactly what the desirable future should be seem to be characteristic of the political and social prophet. If this tendency to moral absolutism is given free rein, it may not only take our thoughts and efforts away from that which is possible, but also help to create a society we could not foresee and do not want.

And these remarks were strongly confirmed by I. I. Rabi's insistence that at the heart of the scientific adventure, at the center of the activity of scientific discovery, there is necessarily an unpredictable element. Indeed, the essence of the modern scientific effort has had to come to grips with apparent unpredictability, with uncertainty at the heart of things, even if statistical generalizations do provide the possibility of fruitful theory. But if the scientist knows that he cannot, in principle, avoid indeterminacy, this does not cast him into a state of hopelessness and inactivity. His work lies before him, discovery and fruitful application are in wait for him; the unexpected relationship, the successful experiment, the chance remark that leads to illumination, the beauty of the sudden emergence of a fuller theory which integrates what previously seemed separate —all these constitute *his* environment of change.

If the scientist's work is to be successful he dare not be intimidated by change, and if his work is to relate properly to mankind he dare not ignore its social implications. It was not a matter of accident that the most dramatic expressions of apprehension and concern about scientific achievements too subservient to military control were to come from the two individuals at the conference who were closest to the world of science, Professor Rabi and Dr. Bronowski. Their warnings were somber, and their sense that human achievement—the immense body of artistic and scientific work, indeed humanity itself—lay under a most threatening cloud was profoundly disturbing.

For some of us there was an even more disturbing prospect, that voiced by Professor Eiseley in his vision of the homogeneous world—the great beast—where human evolution to all practical purposes would have come to an end because all the bright and varied cultures that have in the past made up the life of mankind would have been reduced to one overwhelming culture, no longer really alive, because no longer stimulated by a ceaseless flow of cross-cultural influences.

But we did not emerge from the days we spent together either overwhelmed by a sense of despair and ineffectuality or invigorated with new hope and a sense of mission. Rather what was left for many of us was the feeling that a dialogue had begun, that it was possible to erect bridges across the gap between the "great world" of affairs and the usually tranquil world of the university. The bridges may have been fragile and temporary, but we suspect that today it might be even more possible to erect such bridges. The university is clearly less an oasis of tranquillity than it was when these conferences took place, and the world of business has, perhaps, come to appreciate much more keenly just how intimately dependent is its own health and creativity on the proper flow of ideas and values between these two worlds. After all, they cannot be really separate, and this volume is not an inaccurate picture of what took place between two worlds at these conferences.*

Aaron W. Warner

Dean Morse

Thomas E. Cooney

New York City

January, 1969

* The editors express their appreciation to Mario Salvadori, who, at the request of the chairman, gave an unscheduled paper on "Abstraction in Art and Science" at the Airlie House Conference. This paper had been given previously at a meeting of the Columbia University Seminar on Technology and Social Change and will be reproduced in a volume of proceedings of that Seminar.

CONTENTS

The Environment of Change

The Hazards of Social Revolution

Summary of Remarks of SIR ISAIAH BERLIN

EVERYONE WANTS TO KNOW the answers to the great questions that preoccupy human life: what kind of world do we live in? what is wrong with it? how can we change it for the better? But there is something very peculiar about the fact that although thinkers—both learned men and politicians—have applied all the resources of reason and observation to these questions, history at times fails to turn in the directions they predict or aim for.

Consider the French Revolution. No event was more consciously awaited. Throughout the eighteenth century people expected an overturn, discussed it, wrote books about it, wondered what would make it. By the 1770s and 1780s the atmosphere was charged with impending change. Even though what happened in 1789–91 came suddenly, those who stood to lose most, the aristocracy, the church, the royal bureaucracy were, some of them, filled long before with the expectation of change, even of doom.

That period marks the rise of science and the rise of faith in technology. People supposed in the eighteenth century, partly under the influence of the unique discoveries of Newton, and the vast prestige and success of the natural sciences, that for the first time a real transformation of the human consciousness had occurred. Newton's work really did change the world and the field of possibilities. It was thought that if Newton was able in principle to determine the position and the motion of every par-

SIR ISAIAH BERLIN IS PRESIDENT OF WOLFSON COLLEGE, OXFORD UNIVERSITY.

ticle in the universe by deduction from the relatively few laws
that his genius, and the genius of others, had discovered, there
was no reason why the whole of human life could not be organized in the same fashion. After all, men were three-dimensional
objects in space and time, they were bodies controlled by physical, physiological, and psychological laws. If we could learn the
laws of physics, chemistry, psychology, and physiology, why
not the laws governing human society too? As Condorcet suggested, we study the societies of bees and beavers; and if we can
discover what causes bees and beavers to be and behave as they
do, there is no reason why we should not be able to do the same
for human beings.

The conviction that we should study human society scientifically led to a simple view: the main problems of human life
resulted from various human errors made in the past. These
errors were a result of stupidity or idleness or were the fault of a
minority of wicked men who had deliberately thrown dust in
the eyes of the majority of simple men to acquire power over
them—a lot of knaves leading a lot of fools by the nose. Some of
these knaves had been taken in by their own propaganda; others
knew what they were doing. But in both cases, wicked and ruth-
less kings, commanders, and priests, had misled humanity for
generations. Now all that had to be done was to eliminate the
errors, to expose such propositions as that God exists, that the
soul is immortal, that the earth is flat. These could be disproved
by observation and argument, and once this was done, the task
would become relatively simple. First, it had to be discovered
what human beings were like, what the laws were under which
they operated. Then it would be necessary to discover what
human beings wanted. This could be done by careful physiological, and sociological inquiry. Men were not so very different
from each other; these differences could be noted. It would be
necessary to discover what human desires were, and what means
were available to realize them. All this could be done by dis-

interested scientists. What was needed was the verified truth. Since all the evils in the world were caused solely by error, often "interested error" as Holbech called it, it could be supposed that once the facts were known, once it was discovered what human beings were really like and what they wanted, once means had been found by human genius to provide what they wanted, nothing more would be necessary.

Condorcet expressed this most forcibly when he said, "Virtue, knowledge and happiness are bound as by an indissoluble chain." This led to all kinds of corollaries in the eighteenth century; for instance, that a good chemist is necessarily a good man. It could not be otherwise, because a good chemist is a man dedicated to truth. The truth is obtainable in human relations too. Once found, it was impossible not to live and act in its light. Hence a good chemist or a good physicist or a good anthropologist cannot fail to be a good father, a good citizen, a good man. All human problems could be solved by education; that made men want to search for the truth—and alter nature accordingly.

This was the great principle that animated the French Revolution. The ideal was, of course, to achieve universal justice, universal peace, universal happiness, universal wisdom, and universal fraternity. The whole movement of 1789 and 1790 was one of the high moments of human exaltation and enthusiasm. When people called themselves good patriots in France, they were not nationalistic. At that stage, what they meant was, "How marvellous to live under a constitution, which at last is rational as well as human; which at last has destroyed the privilege of caste and class; which is founded on rational principles and truth, objective truth, which any intelligent man, using the proper instruments can discover for himself; how wonderful that the nation which is achieving this is my own." The ideal was the scientific organization of life in accordance with publicly communicable principles.

But how very different were the results of the Revolution

from those planned! The assumptions of the revolutionists were that man is capable of rationality and of organization: that man, if properly taught by qualified experts, could organize his life on true and harmonious principles. The reformers believed in at least three propositions. They believed, first of all, that all serious questions have answers that can be discovered by human reason. If a question cannot be answered, it is not a genuine question. But if a question is so formulated that, in principle, there must be an answer, then men can and will find the answer. One can, in principle, tell what the center of the moon is like. One can, in principle, tell how various kinds of human beings behave in all conceivable sets of circumstances. The second proposition was that the truth can be discovered by communicable techniques, not by intuitions, revelation, in sacred books, but by techniques that a man of genius can discover and any technician in a laboratory can henceforth apply.

The third proposition was that all the answers are compatible; they can be put together. This followed from a principle rightly found in books on logic, that one true proposition cannot be incompatible with another true proposition. If all questions could be properly formulated and answered, and the answers then put together, the jigsaw puzzle would be solved, and we would know how to make humanity happy forever.

And value judgments? Answers to such questions as, what do we live for, what is good, what is right, why do it, what is honor, why not betray one's friends, why should freedom be preferred to tyranny, why is kindness better than cruelty, peace better than aggression—the truth on all these matters would be established by the application of techniques very similar to those that worked in physics, biology, psychology, economics. Everything that was called philosophical, theological, or psychological, could be converted into positive science. That is more or less what Hume believed. That is what the French revolutionaries believed, what Comte believed, what H. G. Wells believed, what quite a lot of people still believe.

What actually happened, as a result of the French Revolution? The events that occurred after 1789 drew attention not so much to human wisdom, human virtue, human rationality, human organization, human order, but to the opposite. They showed the power of mobs, crowds, and the masses—the exact opposite of what had been planned. They showed the enormous, astonishing influence of great men, of charismatic leaders such as Danton and Robespierre; and they uncovered the irrational forces to which such leaders appealed; they showed the role of accident and of violence in human affairs. By 1815, when the French Revolution was said to have completed itself, the picture is unlike what was planned. Instead of a rational, well-organized, international order, governed by sage assemblies, people were in the grip of acute nationalism—irrational and divisive passions which the Revolution was against. There was not much real nationalism in the eighteenth century. It was thought to be an irrational, emotional drive that placed local interests before those of mankind. It was, in particular, much disapproved of by Frederick the Great, by the Emperor Joseph II, by the French *philosophes*, by Condorcet, by Robespierre. But the French Revolution led to the invasion of Germany, Italy, the Low Countries, and to an extremely violent nationalistic reaction against the French invaders. The result was an outburst of wounded national feeling that is really the source of European nationalism in the nineteenth century. In fact, the word "chauvinism" dates from the French Revolution.

The fate of the French Revolution made a number of people ask why it had gone wrong. The liberals held that it was because the mob had not been sufficiently controlled. The socialists maintained that the economic and social factors lying beneath the surface had not been taken into consideration, and that too much attention had been paid to purely political organization. Historicists Burke, Herder, and Hegel contend that the factor of historical growth—the unquantifiable, qualitative change of individuals and groups, of thought and action—had been ignored or grossly oversimplified. Catholic ultramontanes took the line that the whole rationalist approach was a mistake, that human beings were weak, sinful, and cruel, and that only by throwing onself upon the mercy of God and trusting oneself to the authorized interpreters of the inscrutable will of God, could humanity survive at all.

Later came the more complex doctrine of Karl Marx. He certainly supposed that he was a truly scientific observer of human affairs. Marx's critique of the French Revolution was that the class struggle had not been taken into consideration. So long as there was such a thing as class war and its existence was not recognized, the judgments of excellent scientists, rationalists and ardent improvers of mankind, equally those who did not take enough notice of historical change, would be perverted by class interests, usually because these good people came from a class that stood to win by preserving the *status quo* in certain ways.

This was still an exceedingly rationalist answer. Marx in effect said, "These people have only taken factors A, B, C, and D into consideration; we scientific socialists recognize factors E, F, G, and H. And if we take them into consideration, we will comprehend reality; this will do the job; the trick will be done." The form of the argument was still the same: perfection is possible, humanity can be saved, a stable and decent free society can be created—if only we take into account certain important but hitherto overlooked factors: the economic "base," the class struggle, and so on.

I cannot dwell long on Marx; only note this time-worn fact: that even in his case, history did not proceed according to his prophecies. Marx supposed—with good reason according to his own premises—that in the most highly industrialized countries, in the course of the sheer growth of productive forces, a proletariat would grow, organized and disciplined by the capitalists themselves. But once organized, the proletariat would, if they understood their own interests and power, be capable of shaking off the capitalists, who would become progressively fewer because they were compelled to cut each other's throats in savage competition. Because the capitalists would do so much to eliminate themselves, Marx warned against premature rebellions, terrorism, or disorganized activity of any kind. He recommended that his followers form solid political parties, which would make the revolution by gradually squeezing the bourgeoisie to death.

As we know, very nearly the opposite occurred. The only country in which the program was followed faithfully was Germany. There, a splendidly organized socialist party came into existence. Industrialism increased in all parts of Germany, particularly in Prussia and the Rhineland. The workers grew in number. Their party grew more and more powerful. It built its own world of schools, hospitals, insurance schemes, theaters, playgrounds, concerts; the party looked after its members very humanely; it won more and more seats in the Reichstag. And after the defeat of 1918, the Socialist Party became the most powerful single party in Germany, and ended by being easily and utterly eliminated by the Nazis. This happened in part at least because the Socialists were so beautifully organized, so respectable, so integrated into German society, because they had created so many stable institutions, enjoyed so much security. When it came to the point, they were peace-loving, security-minded, with no revolutionary zeal, no aggressive feelings.

The only countries in which Marxist revolutions of a serious kind broke out, at least before World War II, were the two in which industrialism was least advanced, and in which, according to Marx's principles, there should not have existed an adequate base for political activities by the workers: Russia and Spain. In Russia where there was continuing oppression that was both corrupt and inefficient, one could only operate by the method of *putsches*, small revolts, in the end, organized revolution. This worked. But it worked against the program Marx enunciated. The German Social Democrats obeyed the precepts of the Master blindly and were crushed by the Nazis; the Russians deviated from them and, by doing so, won.

Let us look at the Russian Revolution of 1917. In theory, this Revolution was to bring about a rational organization of society. It was supposed to permit the gradual application of scientific method to the whole of society by means of a proletariat enlightened enough to support it. But again the results were the opposite. The effect of the Revolution drew attention to the existence of forces that had been ignored; it showed that terrorism paid, that charismatic leaders were obeyed, that democratic methods could be ignored, and that minorities could sit on top of majorities and dominate and oppress them in their own name, without effective opposition.

The results of the last two great wars are not dissimilar. The 1914–18 war was fought, among other reasons, for the principle of self-determination. Empires were crushed. Small nations were established and given their proper frontiers. Again the old dogma was applied; here is the disease, and if the doctor is well-enough informed, he has the cure. This is exactly what Wilson thought. But instead of the result anticipated, an appalling degree of economic instability, violence, and general chaos resulted. As Trotsky said, the Treaty of Versailles converted Europe into a lunatic asylum but failed to provide the inmates with straitjackets. World War II was intended to produce some kind of democratic system when Nazism and Fascism were crushed. But again, its results were paradoxical. Instead of ushering in victory by the forces of democracy over totalitarianism, irrationalism, and antihumanism, it ushered in a reign of military dictatorships. There are probably more military dictators and more charismatic leaders in the world today than there have ever been before in human history.

But perhaps the most significant unforeseen result of World War II was to show the power of totalitarianism to organize human irrationality, to play on other people's nerves and aggressive instincts, to utilize everything the rationalists had always denounced. The use of this power by manipulators such as Goebbels was one thing not predicted by any of the prophets in the nineteenth century.

In fact, the experience of the twentieth century showed that an alliance could exist between science and irrationality. This indeed was something new. The general assumption had been that a scientist was a rational man, and would surely not have anything to do with somebody who was the opposite. But we have recently seen the emergence of politically irrational systems (Fascism and Nazism) that used the latest technological devices. And the scientists who submitted to the requirements of these governments felt that they could not simply opt out of the social system. They would do a good job on lines which the state commanded; they would be as loyal to the irrational state as they had been to any other. The combination of extreme competence among scientists, with the violent irrationality of the leaders of the state was something that had never been foreseen. This was a new combination and struck humanity with amazement and horror. And this is again something we learned from experience and not through scientific prediction or analysis of human character.

This terrifying alliance caused people to wonder whether it was enough to say that more knowledge would of itself make our lives more charitable, more humane, richer in esthetic and ethical experience. No people had produced more erudite and brilliant scholars or more magnificent achievements in natural science than the German universities in the late nineteenth and early twentieth centuries. No country had a better system of public education, or had more distinguished historians and philosophers than Germany in the 1920s. By ordinary criteria, it was one of the most civilized nations in the world. Yet, this did not prevent that happening which did happen.

And now, twenty years after Germany's fall, there are still people with ready explanations of our problems. They tend to say, "If only we could get rid of one more obstacle, we really would be on the way to progress. If only we could get rid of the fanatical Chinese Communists (let us say), who hate the Western world or are possessed by some kind of intolerant Marxism, that would rid us of the last obstacle. Or, if we could only get some kind of permanent agreement with Russia, *that* would do it; only the absence of this obstructs our path."

It is the ancient fallacy again. Men say, "We know what the true ideal is, we know exactly where we want to arrive, but the enemy simply cannot be reasoned with; somehow our opponents have been brainwashed. For God's sake, can't this last obstacle be removed? Can't we have a peaceful world at last? If only we could overcome X or Y or Z—Communists, nationalists, imperialists, Americans, Chinese—there are a great many of them; well, that makes the task bigger. . . ."

The temptation is obvious. There are many people who are convinced that they know what is right and good. They know that there seems to be only one final appalling obstacle. If it could only be shoved out of the way, even at the cost of a certain number of human lives, then at last mankind could progress.

By now it should be apparent that there is something wrong with this approach. Whenever it has been used, it has produced not what was anticipated, but the opposite. This is not to say that the approach is ineffective. On the contrary, it has at times had a considerable effect. The wars of religion, the French Revolution, certainly changed things radically. So did Marxism, so did the Russian Revolution, so did World War I and World

War II. Man is not weak or ineffective in the face of forces of nature and history, he is strong, exceedingly strong. When people really do get together and make a revolution or a war, they do alter history, and they alter it in very violent ways, but often not in the ways they anticipate.

These facts suggest two central propositions. The first is that there may be something wrong with the policy that all political questions can be answered in definitive ways. In physics and mathematics and physiology we have no doubt made progress; we do know more than our predecessors, and can prove it. But in human affairs this is more doubtful. We do our best to perform rational analysis, but lessons of experience induce a certain modesty in claims of definite advance.

The second point is that it may not be true that all the answers—if we can find them—are necessarily compatible. This has been the premise of the proposition that final, single solutions to the great social problems can be discovered.

To the French revolutionary, for instance, all moral and political issues—e.g., questions involving choice between peace and honor, liberty and efficiency, equality and liberty, must, in fact, have final and mutually compatible answers. The end is given—we know, said Saint-Simon, the oasis toward which we are all marching. The only problem is how to get the human caravan there. Thus, all moral problems become questions of the correct technical means. One good end cannot (it is thought) conflict with any other good end. The jigsaw puzzle can be solved.

But if we consider some of these propositions, we will see that certain human values are not so evidently compatible with certain other human values. For example, where we say that we want the truth at all costs (if we really mean at all costs), we find that the costs are at times very high indeed. If we say we want to establish and proclaim the truth in all human matters at all times, we will have to tell the truth to a lot of people, some of whom will be deeply wounded by it. How much good will it do to tell the stupid, the ugly, the unattractive, the incurably sick, that they are so? It is not self-evident that truth and happiness are always compatible—although this seemed so to the people of the eighteenth century. Neither is it self-evident that the highest efficiency and the widest liberty are necessarily compatible. Obviously, to allow unimpeded technological advance, human beings have to be organized in all kinds of ways which they may not understand, may dislike even if they do understand them, and may often resist. The questions are: How far can one justifiably go in trying to break their resistance? What about human rights? If we really believe in the supreme importance of technological organization (as Saint-Simon did), then we must (as he did) deny the existence of human rights; it would be argued that no man should be allowed to resist the rational organization of mankind without which there can be no human progress, creativity, or happiness. The people who refuse to obey well-laid plans for their happiness, and to move to their new houses (jobs, countries) are enemies of progress; we may be sorry for them, but they must be made to move. Rational organization is guaranteed to make humanity happy; we cannot stop and consider the prejudices or the sensibilities of a few individuals who happen to cling in a sentimental fashion to some irrational or traditional form of life. Where the reward is so vast, all obstacles must be swept away and the price paid.

This leads some people to feel qualms. In *The Brothers Karamazov*, Ivan Karamazov says that if he is told that human happiness can only be obtained by the torture of an innocent child, his answer would be, "This is not a price I am prepared to pay. If that is the cost I have no wish to see the performance. I return the ticket." How many innocent victims are worth sacrificing for how much human happiness? Hundreds? Some would say yes. Thousands? Some would still say yes. Hundreds of sue them because they embody the elimination of the last great obstacles, after which the gates of paradise will open. Evil things must be attacked and good ones promoted. But not because the end of the struggle is at last in sight.

Two further reflections. First, some good things appear to have a darker side. In our own age, for example, we point out that automation has liberated people from all kinds of degrading forms of toil, that as a result there is a much wider and nobler future before them. But when we produce more leisure we also tend to produce the possibility of boredom. There is no doubt, for example, that boredom has affected young people in my own country today. The young men of England do not greatly fear unemployment or poverty. Most of them have jobs, and many have acquired material possessions which their parents could not afford and their grandparents could not even dream of. Having got these things, some among them naturally ask, what next? Having more leisure on their hands and feeling uncertain about how to use it they become exceedingly exhibitionistic, and now we have the young men in Carnaby Street suits whose unusual appearance is some sort of protest against society as it is. This is not a form of political indignation against specific evils, as some people try to maintain, but a result of the fact that, somehow, these young men feel stranded—without direction or goals to which they feel committed. This is what some of Mr. Osborne's plays are about. His heroes feel that, idiotic as the old rigid ways of loyalty to king and country, or service in the army and a traditional family life may have been, nothing has taken their place. The army may have meant fighting a lot of Indians or Africans, and loyalty to the family may have meant oppression by some tyrannical head of it, but still these were stable forms of life—there may have been misery and injustice, but the framework was firm.

Obviously we should not refuse to liberate prisoners simply because they do not quite know what to do with their freedom.

thousands? Millions? Some would stop only at that p
even there.

The fact that such figures can be considered by :
lators, while others reject the very notion of such
horror, indicates that there is some kind of collisio:
here, and that it is not clear how it can be resolved.

The same can be said about liberty. If every
really free, the stronger could hit the weaker on the h
say, "Yes, I'm afraid you'll have to forego some of yo
because others are human beings too, and they too h
they have the same rights as you; in fact they are yo
The liberty of the pike is not compatible with that c

The general assumption, then, that all values can
fortably side by side; that we could have perfect free
perfect equality, with perfect efficiency, if only we
to achieve this, is not a self-evident truth at all. It w
rather, that certain values are not merely *de facto* in
with certain other values because we are ignorant, or h
only have limited resources to be divided between th
is also, apparently, an inherent conflict; there seems tc
superable conceptual incompatibility between perfe
and perfect mercy, between perfect equality and perfe
All these are human values, and if anyone is ready to c
of them, we respect him for it. But how do we decide
these various values to pursue, or how much of wh
should be sacrificed to what other value? It is not easy
what guarantees moral progress or what *is* the unique,
of men on earth.

Of course, this is not an argument for doing no
fact that choices can be agonizing does not entail that
not to make any choice at all. But there is a vast diffe
tween saying that ends are worth pursuing for their c
because they are good or right, and saying that we sh

But we must recognize that a certain kind of liberation creates its own problems.

The second reflection is that good is at times achieved in unexpected ways. When we examine the history of human happiness, toleration, peace, of all those ideals to which we in the Western world are wedded, we realize that they have not often come about as a benefit of the consciously thought-out, rational application of universal plans made by infinitely wise social engineers or technologists. Mostly they have been attained the hard way. If we ask ourselves, for example, how religious toleration came to the Western world, we find that it came only after the Protestants and Catholics had fought bloody wars to the point of exhaustion in the sixteenth and seventeenth centuries. They fought until enough of them realized that if they went on fighting they might be destroyed themselves. And although neither side gave up its principles, and each thought the other damned, they stopped fighting because they were afraid of a Pyrrhic victory. It was not the preaching of wise men like Erasmus or Comenius, or a general trend toward toleration, rationality, or peace that influenced them, but simply the problem of survival. Because they exhausted each other in these battles, they finally stopped fighting, and toleration became a kind of *de facto* compromise, or armistice, that lasted, fortunately, for quite a long time. And in this way many other human blessings have been achieved. People start with a fanatical belief in the possibility of some monistic solution, some kind of single end which, once achieved, will ensure happiness to mankind. They believe, therefore, that anything is worth sacrificing, because *the* solution—whatever it is—will surely be the *final* solution. Their belief in a monistic solution tends, in the end, to be toned down to a compromise. We should live to hope that this is the worst of the argument, of looking at the facts of human history, of the advances of psychology or sociology. But in fact, compromises are usually reached only after various factions fight and fight until they can no more, and the dreadful cost gradually leads men to the realization that they must learn to tolerate each other. It is very difficult to get people to tolerate each other unless they have tried intolerance and failed. This is a melancholy reflection on human character.

What should we do? No doubt we need more knowledge of ourselves—individual and collective, psychological and social. No doubt we must admit that we have not studied ourselves enough; there are all kinds of dark and irrational "drives" in us that have not yet been properly understood. The French reactionary philosophers of the late eighteenth and early nineteenth centuries, influenced perhaps by the loss of life in the Revolution and the wars that followed, claimed that what people really like is not cooperation, but collective self-immolation on a common altar. Armies are told to march, and though they do not know where or why they are marching, they march. Mutiny in an army is comparatively rare. What people evidently adore, one of these pessimistic reactionaries said, is being slaughtered together for some unintelligible ideal. This is exaggerated, but so is the optimistic opposite. Too many rationalists have not noticed these dark destructive forces, or do not consider them or accept them until they hit their own heads against them. By that time it can be too late. They begin by discovering the solution, the official solution, and they must embrace it—the simpler it is, the more attractive. That is why the prophecies of the system builders, in fact, often come to naught.

Contemplation of modern history alone ought to lead us to the view that there is no simple solution, such as that the answer lies in technological advance, or a return to some ancient faith, or self-determination, or utilitarianism, or Communism, or Socialism, or individualism, or capitalism, or any other ism or single idea. If our generation can learn anything from the past, it ought to contemplate two suppositions about human goals. The first is that not all goals are compatible. The goal is chosen for itself, not in the hope that success in attaining it will guarantee the attainment of other corollary goals, but with the realization that it may foreclose them. If X is chosen, Y may be irretrievably lost, and nothing can be done about it.

The second supposition is that human beings, in the process of seeking their goals, transform themselves. And by transforming themselves, they alter their goals too. We begin by seeking goal A, let us say economic prosperity, or social equality; and the more we improve our social structure and outlook and behavior, the more our aims and goals will in turn alter. We cannot therefore predict today what our goal will be tomorrow. Any attempt to put humanity in a straitjacket, no matter how noble the intention, is dangerous. We say that what men really want is A, B, and C, so we will provide them with A, B, and C, though of course at some cost. One must (we are told) break eggs to make omelets. Very well, we break the eggs—but the question is whether the omelet has been worth it. After breaking the eggs, even though the omelet is edible, those for whom it is made may no longer want to eat it. Their very success has altered them; they now want quite a different dish. So more eggs have to be broken, and so the egg-breaking process can continue forever, so long as we insist that a given omelet is the final good of our—or mankind's—desire. It is better to be more scrupulous about breaking eggs—or lives—once we know that no matter how many are broken, the dish produced may well not be what, by then, we shall heed and idealize.

These two suppositions seem to constitute one of the few dependable lessons of history. We cannot fully predict the future, and we have to realize the necessity for choosing among incompatible ideals—and for living in a society in which different people may seek different, equally valid, ends. We must learn to be satisfied with the maximum effort to preserve some kind of precarious equilibrium between varieties of goals and of men, a system of world order in which necessary change can occur without breaking the delicate crust without which human life cannot exist, and without which no ideals can be properly worked for and preserved.

This is a very difficult and a very undramatic thing to do; it can be tedious, and the tension is not good for the nerves. It is a great strain to have continuously to be aware of the fact that all we are trying to do is preserve a social framework or individual health which is constantly threatening to crack and requires to be patched and propped and protected. It is far more agreeable and exciting to have a shining ideal, to think that we are approaching it because we have already conquered a given number of obstacles, and there are only a finite number more, after which we or our children or grandchildren will enjoy perpetual sunshine. The most noble and moving of democrats, Condorcet, was sure that the day would arrive at last; the day on which mankind will be happy, free, and wise. But this is not very likely, because it is an *a priori* truth that one cannot have everything. It was said long ago that we cannot recapture that which made the ancient Greeks or Jews or Indians wise or happy or great. That is gone forever.

This truism—that we build for our time, and then we shall see—is the strongest argument for what must be called a rather untidy liberalism. Values are not less sacred because they are not eternal. This is a liberalism in which one is not overexcited by any solution, claimed finality, or any single answer; where, above all, one is not deluded by the thought that one is called upon to remove terrible obstacles that are the last great stones that stand before the doors of perfection; and that the destruction of entire societies is not too high a price to pay for victory in the war to end all wars, the overcoming of the last great obstacle, after which prehistory ends and true history begins.

OPEN DISCUSSION

QUESTION. Is there a positive side to your view of the dynamism of social change? Or are we bound to get into trouble because things will always turn out to be going in directions orthogonal to the ones we want to take? We Americans are committed to action, though, and we're not going to stop trying to change. What guides to action can we derive from the past?

BERLIN. All I want you to do is to be wary. I do not want people to feel that we can go forward in any direction we like. The path of true wisdom is to act with a maximum awareness of the consequences.

What I am protesting against is those starry-eyed individuals who act with blinkers on. This seems to me to be one form of the modern psychosis. All I am proposing is that when we see things that are not good, or should be improved, we act, but that we do not assume that we have all the bricks, so to speak, in the one single splendid edifice that might be erected. What seems very important to me these days is that we must understand that there is no such thing as one way to see the world, one path for mankind. I think that way really does lead, inevitably, to fanaticism and also to great hate without qualms.

I do not think, however, that the only alternatives are torpid inaction or cynical despair on the one hand, and, on the other hand, ranting at or crushing our opponents in a huge optimistic and aggressive drive for perfection. We must always remember that the consequences may be different from what has been planned. We must always be prepared to patch up, to remedy, and we must not be afraid of doing something even though it does not have glittering perfection. I am very sorry if one does pick up a pessimistic and rather anti-activist view from what I have said. I want to repeat again that, on the contrary, I think that the danger of subsequent inactivity or of some kind of disillusionment is greater if we try to do too much than if we believe that we must proceed on a piecemeal basis.

Take the Russian experience. It is true to say, I think, that the whole Russian transformation is really a strange story of gradual infusion of water into the original Leninist wine. Lenin initially—and quite genuinely—believed that the Russian Revolution would have been a futile act unless it was a prelude to world revolution. And even trying to achieve Socialism in one country, which was Stalin's solution, already represented a certain injection of realism into what otherwise would have been an absolutely intolerable situation; the Russians would have had to go on fighting the whole world forever. The Russian Revolution, quite apart from its totalitarian aspects, quite possibly has on the whole been going in the direction of greater and greater realism.

But the degree of realism is created by the emotions of the time—we have to admit that. And here we must go back to the lesson of the Protestant–Catholic wars of the seventeenth century. Because the Chinese are much more fanatical than the Russians, the Russians are appreciably more reasonable. Everyone knows that they are far less fanatical today, not because they have convinced themselves by intellectual argument of the value of reasonableness, but because they simply wish to avoid a head-on collision with a power more fanatical than they are. They wish to have a world in which it is possible to coexist with the Chinese, just as Catholics and Protestants wished to see a world in the seventeenth century in which they wouldn't all perish. This seems to be the most powerful factor, on the whole, in the direction of moderation, humanity, reason, and *ad hoc* progress. I do not see any reason why this shouldn't ultimately result in a world in which, though a minimum of friction will always exist, we will, nevertheless, not be in danger of mutual extermination.

QUESTION. I suppose no one here would disagree with the philosophy Professor Berlin outlined; but if we look back at the process of change, let's say even in modern history, in the last few hundred years, do you think that the formula you're prescribing would have been adequate, or compatible, with what happened? Just take two institutions referred to. Do you think if Jean Monnet had started out with a kind of low-key presentation of his notion of the Common Market, or if he had not exaggerated what it would do, the innovative character of it, he could have accomplished what he did? Or what would have happened if Roosevelt, Churchill, and Stalin had said of the United Nations that it would only be a place where people could talk? Can large human movements be mobilized if people know in advance that the project is half-defeated on some things? Isn't ideology such a necessary part of the law of social change, of social movement, that you cannot dispense with it?

BERLIN. I agree with you. I think it is a very telling and searching criticism of what I said. You're absolutely right. Nobody ever gets his point of view accepted without exaggeration. Unless one asks too much, one doesn't get a quarter of what one wants. And, therefore, there is in effect a dilemma here; you are quite right. For example, take another episode involving Monnet, before the Common Market. When Monnet arrived in this country he, more than any other man, I suppose, was responsible for the raising of the sights of the American defense program by subtly suggesting figures of aircraft production that seemed absolutely fantastic to the generals and admirals of that time. I think it was only because the imaginations of key members of the government were caught by this that the United States government came to adopt the goals it did. So that in a sense you're perfectly right. Unless one has bold vision, has very large ideas, one cannot go forward at all. At the same time, the sort of people who have bold vision and large ideas cannot be psychologically the same sort of people who say, "I'm asking something enormous, but of course I know I shall only be getting the tenth part of it." Once you become conscious of the artificiality of what you demand, it robs your argument of that necessary to ignite other people.

COMMENT. One cannot do without ideology, really.

BERLIN. I think that progress occurs, by and large, under conditions of exaggerated claims, if you like. All I wish to say is that these claims are exaggerated, and that when they turn out to be exaggerated, people shouldn't be too surprised. But, of course, one cannot expect them not to be too surprised, because if people were not going to be too surprised, they would have seen the exaggeration at once and would not have acted upon the claims. How we can get out of this dilemma I do not undertake to say.

QUESTION. I think the fact that the results of our objectives are usually different from what we set out to do creates the incentive in us to face new problems in the future. And if we don't have problems to overcome, we're not going to advance. Perhaps, if we were ever to reach this utopia that has been alluded to, we would find it very boring and very uninteresting, and it would create the greatest of all problems. I feel that this is an optimistic view of the future. Did I misunderstand?

BERLIN. No, you have understood very well. I quite agree with you. I neglected to say that all those young men in England whose immediate problems have been solved now have largely one problem, that of boredom—they do not quite know what to do with themselves. Really, by the standards of their grandparents, they ought to be very happy. If we do not have change, then we have lost our incentive. If we look at the ancient world and ask what made Greek philosophy decay, the answer is, according to Hegel, probably boredom. At the end of a stage in its career it became ossified; nothing further happened.

The Roman Empire certainly collapsed as the result of a kind of boredom. The number of barbarians attacking Rome in the fifth century wasn't greater than the number of barbarians attacking it in the third. But the Romans had a very stable form of government that was regarded by the upper classes as a perfect form of government in many ways. Bureaucrats bred bureaucrats, and gradually the whole society did exactly what their grandfathers had done. The roads slowly fell into disrepair, taxes grew, armies became larger and larger—this is what actually happened but nobody quite knew what it was all about, because there were no positive objectives. All they did was to keep barbarians out and to continue living their dreary lives. In the end, if we look at the writings of late Roman historians, we detect that they didn't quite know what they were defending, what they were for or against. The world had become stale and unprofitable. You are right, not only is the utopia unobtainable, but it would be appalling if it were ever attained.

QUESTION. I am interested in your proposition about the incompatibility of values. Within ourselves and within even our states, we probably have the ability to cope with these conflicts. But what about the problem of interstate—or international—incompatibilities? Do you find the United Nations adequate to deal with them?

BERLIN. No. Nobody finds the United Nations altogether adequate. Nevertheless, it is half a loaf and a great deal better than nothing. What the U.N. is accused of is ineffectiveness here and ineffectiveness there. It is accused of being what the old League of Nations was—a talking shop. But there's a great deal to be said for talking shops! A fearful amount of time is wasted at the U.N., and widespread hypocrisy, a vast amount of bromides, an enormous number of insincere platitudes prevail there. This can't be helped. An enormous number of pointless speeches are made by people who don't believe them to people who don't listen. Even so, all is well as long as people talk— Winston Churchill once said he preferred a "jaw war" to a "war war."

Even if the U.N. provided simply the opportunity for the African leaders to sound off and have their grand speeches, reported in glowing terms in their countries, rather like members of our Parliament in England and your senators, it would be a good idea. It makes for a certain degree of cooperation, of non-friction; it prevents to some degree warlike activities.

QUESTION. You cited the dangers that result when science places itself at the disposal of an irrational state. Isn't there some merit to the view that the use of the scientific method is ultimately incompatible with the pursuit of irrational ends, and that given this ultimacy, there is some pragmatic consequence, too?

BERLIN. I do not believe that the scientific method is ultimately incompatible with irrational ends. I wish I did. That is what was believed in the eighteenth century. If they were right, then progress would be enough to save us. But this seems to me unduly optimistic. The question of what kind of goals we seek is not easily influenced by the mere presence of scientific method. It seems to me that it is perfectly possible to be fairly constant to the methods of science, when you are dealing with the physical sciences. However, applying scientific criteria to some parts of human life produces the kind of artificial uniformity I have mentioned that, in turn, has the rather unpleasant results we are all familiar with.

DISCUSSION: SIR ISAIAH BERLIN AND I. I. RABI
ON THE "RATIONALITY" OF SCIENCE

Following Professor Berlin's talk, a discussion between him and Professor Rabi arose that was of particular interest as a public confrontation between a Nobel laureate physicist and a humanist of equal scholarly eminence. A participant had asked Professor Berlin whether the rationality of the scientific method was not ultimately incompatible with its irrational use by a society. After Professor Berlin's answer, Professor Rabi began the discussion.

RABI. I'd like to point out some respects in which science is not rational. In fact, science never has been wholly rational because it is involved so centrally with empirical things. The world is an open-ended world, and one cannot be rational about it. There are elements of rationality in science, to be sure, very strong elements of rationality, but to say that scientific activity is rational is to confine ourselves to the history of science and to forget that science is also in the present.

BERLIN. Then what do you mean by the term "rational"?

RABI. The coherent structural quality of logic, of grammar, or a system of law. Things of that sort are rational.

BERLIN. But wouldn't you say that scientific results would not be accepted by a body of scientists unless there was some general criterion for establishing what is and what is not to be accepted?

RABI. The border between the accepted and unaccepted in science is a very hazy one. Quite definite scientific results sometimes are not accepted for a long time for quite irrational reasons, such as habit and prejudice. Sometimes, people will hold to a theory quite irrationally—an example is the theory of phlogiston. Science is a much more human enterprise than people usually appreciate.

BERLIN. May I press this point a little further? I really don't know the facts. I can agree that science has arrived at its results by all sorts of procedures that may not be rational: by hunches, intuitions, guesswork, genius, and so forth. But surely the facts ultimately incorporated in what might be called the tested body of science are only those for which some argument exists, for which some kind of proof can be put forth which, if it is not absolute proof, involves at least some sort of probabilistic argument that fits into a given structure and doesn't clash with what has been accepted before. And this procedure is, of course, a force making for a general notion of order, coherence, and compatibility in science.

RABI. That is why I used the analogy of the law. It takes the new event and places it in an established structure. The event then becomes an established, accepted thing. For example, without any great Ptolemaic theory or Copernican theory, I would say the sun will rise tomorrow. I would even be able to tell you how long the day will be by looking at my references. Most of science that you call rational is in that sense simply established.

But where is the growing point of science? When does one learn new things? Certainly the whole aim of science is to build a structure which enables one to go forward, to learn. The turning points that are the most important are the revolutions that occur in science. Almost every scientific advance, in the sense of a rational understanding, has carried within it the seeds of its own destruction. It is true that science is a great structure, but that structure is just a human creation, a result of our tendency to erect a system of laws. Its foundations are sometimes destroyed, but, strangely, the edifice still stands. This is what I meant when I said scientific activity is not rational. Actually, when you deal with scientists, as you have, Professor Berlin, you find that they are not the most rational people.

BERLIN. True, progress in science comes by revolution in which part of the foundation is destroyed. I concede that science is in a state of continued flux, a state of turbulence. You are surely right in your description of the process of scientific discovery. Nevertheless, isn't it the ideal to incorporate the new into a fairly stable unity with the old? Isn't that the basic purpose?

RABI. The idea is that the world is understandable.

BERLIN. Am I right, then, in assuming that the ideal is to extend our understanding of the world by establishing connections among the different parts of science?

RABI. That is a goal, yes. It is an ideal, and when two separate sections of knowledge are fused, such as physics and chemistry, it is considered a very great step ahead. These two fields grew up quite separately. Even today they are different disciplines; represent different ways of thinking; and have different values, different standards of truth, different standards for conviction. When going from one scientific discipline to another the laborious and difficult task is not learning the facts of the other, but acquiring the spirit of the discipline, its tradition, its ways of thought.

The Revolution in Science

I. I. RABI

I WANT TO TALK about the effects of science and technology on society, and how we may cope with them better than we have so far. What I have to say is based largely on our recent American experience, which seems to me to epitomize the power, the promise, and the dangers of science and technology. Although I speak of dangers, let me make my own position clear to begin with: I am an optimist. I believe that the world has made enormous progress through science and technology. In my own lifetime in this country I have seen that progress in every field. There is greater morality; there is more justice; there is more education; there is greater comfort; there is better health. Our influence, by and large, has been for the good. The American influence has upset tyrannies and empires and has done a great deal to make men free all over the world.

Therefore, I am not here to bemoan our state. But I am greatly worried about the future. We in the United States have grown so powerful economically, militarily, and in many other ways, that we can no longer afford merely to gloat over our power. We must begin to understand it, so that we may control it and use it wisely. But our problem is that these profound changes in our environment seem to be only partly under our conscious control. To a very great extent the scientific and technological revolution seems to be something that is essentially just happening to us. Whatever science does to change the world, it

I. I. RABI IS UNIVERSITY PROFESSOR EMERITUS, COLUMBIA UNIVERSITY.

does in a subterranean way. The scientific attitude is not immanent in our society, even among those who make the decisions. The scientists themselves reflect this quality of the times, because the scientific approach has never been generalized. When a scientist is put in a position of social and political responsibility, he acts like any other man.

Neither our day-to-day decisions nor our basic philosophy reflects the view of the universe and its possibilities that is at the heart of modern physical science. In the closing years of the last century and in all of this century a new and more subtle physics has arisen, the physics of the atomic and subatomic world, a world hidden from the direct apprehension of the senses. This is a world of light quanta, electrons, protons, neutrons, neutrinos, mesons and baryons, and other objects only dimly apprehended.

The main job of physics for the foreseeable future will be to understand this new subnuclear world. It has already been clear for almost half a century that the basic laws of physics are infinitely more subtle and profound than had been imagined by the founding fathers of classical physics. All the basic concepts of time, space, and causality have undergone profound changes and are not yet well understood.

The ideas of science simply have not penetrated very far. The changes we see are the material and circumstantial ones. We do not seem to be consciously trying to understand human life in scientific terms. Do we even consider, for example, whether present values and goals are consistent with what we know or could know about the psychology and physiology of man? Are we doing anything to avoid the death from boredom that so many have forecast when the majority work perhaps fifteen hours a week? Picture Florida dense with elderly and semi-elderly people, caravans of them clogging the roads, the whole state looking like Long Island beaches on a hot summer day.

We also hear a great deal these days about the power of scientists in government and society at large, but actually they have very little real power. The presidents of our great corporations, of our great universities, the heads of our television and radio networks, the people who control our newspapers, the presidents of banks, the leaders in labor unions, the members of Congress, of legislatures, of boards of aldermen—very few of these have any scientific or technological background. The scientific community and the scientific way of thinking are not represented in the councils of power. In fact, the image of science in the minds of most educated laymen is an archaic one. Physical science has been moving ahead for the last fifty years upon principles of relativity and indeterminacy, while the layman's view of science, of the universe, and, as Professor Berlin has pointed out, of society itself, has been based on a Newtonian faith that if we know enough about the present we can predict the future with certainty.

We may say, then, that although we live in an age largely produced by science, we do not live in a scientific age. How have we arrived at this paradoxical situation, and what can we do to remedy it? Let me try to answer the first part of that question with a little history, both personal and general; let me answer the second with a most tentative prescription for a kind of education that will allow people to feel that they are part of the environment while they also realize the necessity for understanding and controlling it.

I have had the good fortune to live in a very interesting period of the development of American science. Let me therefore repeat here some observations on the development of physics in our country that I made a few years ago before the American Historical Association.

To me, it has always been a puzzle why the development of American physics came so late. The establishment, for instance, of Johns Hopkins followed about a hundred years the establishment of this republic, and Johns Hopkins was just one isolated instance. There was no question but that the country was large and wealthy at that time and had very important industries—but, up to then, few research institutions.

Unquestionably, Benjamin Franklin was a great physicist, but he left nothing very much in the way of a school that had any viability. Henry was possibly one of the great physicists of the nineteenth century, or might have been if he had happened to be in the proper place, but he also left no school. Gibbs left no school.

Louis Morris Rutherford of New York was a great astronomer, although largely unknown, and he became a man of means, went abroad, studied astronomy, came back to this country, and made the first photographic telescope with his own hands. He made measuring engines; he understood the future and nature of astronomical instruments. He was not just an amateur taking pictures. But as far as I know, he published no paper, left no students. He left his telescope and his plates to Columbia, and they were the basis of several careers afterwards.

One can go on in this way and show instance after instance of men of high quality who somehow or other could not act as nuclei for the crystallization of a group or a school in which to carry on work of a level comparable to their own. We do not see such schools until the very end of the nineteenth century—in fact, until we get into this century. I have no explanation for it. I think it is an extraordinary social phenomenon when you consider the wealth and size of our country compared to other countries.

In Europe, there were academies. There was the patronage of the wealthy; there was the patronage of the court; there was some centralized way in which science could be supported and stimulated. In this country it had to happen by other means.

Around 1900, the American Physical Society was founded, and Roland, the great professor at Johns Hopkins, was its first president. His presidential address is extremely interesting because he inveighs against the conditions of scientific research in the United States, pointing out how the country worships false practical gods rather than the search for pure knowledge.

The formation of the American Physical Society really marked a great change; we in America had found an institution which was uniquely our kind of institution. The American Physical Society is an entirely voluntary association, supported by the dues its members pay. It was set up to publish a journal, the *Physical Review*, in a typically American way. It was set up in such a way that the editors worked without pay, and the secretary and the treasurer worked without pay; every effort was made to get this journal out at as low a price as possible so as to reach the largest possible number. The Society held regular meetings where the physicists could get together and draw support from one another's presence and papers, and have a forum for the discussion of scientific matters.

This Society almost immediately began to bear fruit. I will skip a period until I can speak from the evidence of my own experience which comes in the mid-1920s. By this time, the Physical Society was well established. It had a few thousand members and the *Physical Review* was a recognized journal, published once a month. There were about four meetings of the Society a year, two of them big ones held in New York and Washington.

All this time the numbers of students who were studying physics in the universities were increasing. If you will excuse me, I will become somewhat personal and describe what I think was a typical situation in the mid-1920s. There was no question but that we had a number of notable physicists in this country—Millikan, Compton, to mention but two. There were a number of institutions which could give courses of instruction in advanced physics which were excellent, given by serious men who knew the subject. There were a number of institutions which had laboratories of fairly good quality. There was research going on at about fifteen or twenty places in the country which would lead to doctorates. The students were very devoted and very hard workers to an extraordinary degree. Sundays and holidays meant nothing. If you walked into many of the better laboratories, you would find them very well populated.

I was a typical product of one of these universities—Columbia. I had done graduate work at Cornell. Then I was able to get a small fellowship which enabled me to go abroad to study and, of course, this was the ambition of every serious student of physics. Although good things were happening in the United States, the great things were happening abroad, particularly in Germany.

When I went to Germany, I expected to find that my education had been rather poor. There were no men at Columbia at that time of the stature of the people in the institutions to which I went for a visit. I expected to find that I was a rather ignorant young physicist. To my surprise, I found that my education and that of my colleagues was far superior to what 95 percent of the German students had. We had had a broader and more rigorous training across the field of physics than those people. We were less specialized and we knew more about more things that the German students—the 95 percent—did not know about. The other 5 percent was really hand-raised, in close association with the great men with whom they were studying, but the average among all students was low compared to the American product of our better universities.

So, I can truly say that as far as knowledge is concerned, I learned nothing in the two years I spent abroad before coming back.

However, in another sense, I learned everything. It was the first time I had come in contact with men of great minds, men who had made major contributions. I was able to measure myself against them and find that physics was not the sort of thing that only they could do, but that I could, too. It rid one a bit of this feeling: Are we Americans creative or are we not? And I think I can speak in this for other Americans of my generation whom I met in Germany: Oppenheimer, Condon, Robertson, and Pawling.

Furthermore, it involved what I would call the acquisition of taste, which I think is most important in any creative effort. It is often just as difficult to do something which is quite uninteresting and unimportant as it is to do something very important and constructive. It is a matter of taste. I believe it comes from environment, although some have more taste than others. I think this was the main thing which we got out of Europe. I think by and large, as I review my colleagues, those who took their doctors' degrees in the United States and went on to Europe did better than those who went to Europe to get a doctor's degree during that period. The American Ph.D.'s went there with their information and they could absorb the atmosphere; whereas, if they went there to get the Ph.D., they were in competition with the top Germans who were working very hard, who were in a very definite routine, and, in that sense, wasting time.

To give you an idea of what happened in a short period, I arrived in Goettingen in 1927 and I found that they subscribed to our journal, the *Physical Review*, but they waited until the end of the year to get the twelve issues at once to save postage. That was the standing of the journal. There were good things in it, but one really did not have to rush. There were the criticisms of American work that we like to make of Japanese work and now to some degree of Russian work. It was unfinished, raw, provincial, not too significant, and so on, with occasional flashes of brilliance, as in Compton's discovery of the Compton Effect.

Ten years later, the *Physical Review* was the leading journal in the world. This was not because Hitler threw out the Jews, although that action certainly enriched our American communities. It was the boys we trained—my generation—the American boys we trained, and the schools we set up. It was our own efforts, and particularly the efforts of our students. For a hundred years, Americans had been going over to Europe to study, but they were few in number, and they came back and somehow or other got jobs here and there, and the world descended on them; they were lost because they were busy, and they had nobody to talk to.

But in my time there were enough who had gone over so that when they came back their number was sufficient to form a "critical mass" (to borrow a term from bomb technology). The people could get together and, of course, there was the American Physical Society with its regular meetings. So, the time was right.

Also, around 1929, the country felt rich and these young men who came back got decent jobs where they could be independent. It is difficult for me to describe the general ignorance about the growing point of physics among the professors in the best universities in the United States.

Almost all my colleagues will agree that one of the great instrumentalities of the change I am talking about was the institution of the National Research Council Fellowships. I think it was through the generosity of John D. Rockefeller. A certain amount of money was made available for fellowships for students, for people who after receiving their doctorates could go on for a year or two of further study either in this country or abroad. This has made an enormous difference to many subsequent careers. Previously a man had to step out into a low-grade teaching position where he was overwhelmed by his teaching chores, where his rank was so low that he could hardly argue with his senior—our hierarchy was much stronger then than it is now—and his spark of interest was soon extinguished; the means were not at hand for further study. The man would not be important enough to get them, and a great deal was lost.

The National Research Council gave him a year or two either here or abroad where he could do research, come back and prove himself, and get a better job, to select an environment which was favorable for research. I think most physicists will agree that the National Research Council Fellowships, although they never amounted to very many, were an important factor in the rapid build-up of the base of American physics.

Europe at that time—and I am still talking of Germany although it was true of France—was already bearing the seeds of decay of its scientific tradition. Once in early 1929 the Theoretical Physics Seminar was invited by the Theoretical Seminar in Berlin to a joint session. There must have been something like a dozen of us there. Out of that dozen there was one German, who was Werner Heisenberg. The rest were American, Chinese, Austrian, Bulgarian—but mostly American. There was not a single German outside of Heisenberg.

At the same time, the United States was producing these very large numbers of people, not of the surpassing excellence of the few Germans, but people who, on the whole, were well trained and with encouragement and the proper environment could really go ahead. So, we had something like an iceberg, a small part projecting of the people whose names one can readily recall of that period, but below that was a large mass of well-trained people. But in England and France they had only a few individuals in the ranks below.

One other very important point which is hard to believe, but nevertheless true, is this: In that period, I was astonished at how much better the experimental equipment available in Germany was than what was available in the United States—and mostly with American money. The Rockefeller Foundation was busy helping universities in Europe, whereas we were very, very poor in our universities as far as equipment was concerned. I had much better equipment to work with in Stern's laboratory in Hamburg than I would have even at Columbia, which was not one of the poorer universities, so that also was an important item.

The period 1930 to 1940 saw an enormous growth in the

United States. I cannot describe what the total impact of the depression was, but one of the effects was that people became more serious and there was perhaps a greater interest in things of the mind than there had been previously. There was not enough money for more material pleasures. Then we began to get some money from a few of the foundations such as the Rockefeller Foundation and the Research Corporation. Money was given to Ernest Lawrence for his first cyclotron. Here and there one could get a few dollars on which to support a post-graduate student as a research assistant, or one could get an important piece of equipment.

While I am on the subject, I will give you an idea of the change of scale that occurred. Before the war, I had about fifteen people working with me at Columbia and used up about half the total research budget of our department, which was about $8,500—my half. The group now is somewhat larger, a few times larger, but we spend $2.5 million. The university has increased its appropriation from $15,000 to $30,000 and can hardly afford more, and the rest is from the government. American science, experimental science, and even theoretical science lives on government support.

I would also like to say a few things about the war period and what happened in American physics. This was a most interesting period which deserves some very profound study as a piece of intellectual history apart from the weaponry, because we did two things. We killed all pure research in the universities. We were told that this is a total war and that we were fighting two enemies who believed in total war. So, in good old American competitive spirit, we went in for total war. It turned out that we became infinitely more total than the Germans or the Japanese. With very few exceptions, we went around to the various laboratories and took every productive scientist from his laboratory; we took scientists from their teaching laboratories, and we put them into large laboratories for the war effort. The

Radiation Laboratory at Cambridge was one of them; another was at Los Alamos, and so on. There they were, in very large numbers, devoted to making weapons, to the problems of war. This was to my knowledge the first great attempt at marrying the military and the scientific.

The result was immediately apparent, and after a short time I felt very thankful that this had not happened earlier in human history, because none of us would have been left alive. We discovered, of course, that it is very easy to kill human beings if you apply your mind to it with the background of science and if you have people who are trained to think unconventionally.

This combination of nonmilitary thinkers and fighting men to apply their innovations, as history will show, turned out to be a tremendously potent force—so much so that in the short time from 1940 to this day, which is twenty-seven years, we have become deeply frightened about the future, not only of the country but of the whole race. This is a new discovery, this application of the most forward-looking science to the military —and I shudder to think of the next twenty-seven years. Something has happened which is irreversible, and we are facing a monster we must learn to control—and quickly.

It has occasionally been suggested that the way to cope with the pace and unpredictability of social change is to declare a moratorium on the development of science and technology. In a sense I sympathize with this view, and I feel that we might even be able to bring about something like a moratorium. For science can be stopped. Science was flourishing in Italy at the time of Galileo, and the Church halted its progress. It did not revive again really until the time of Volta, two hundred years later. As a result, Italy became a rather unimportant place for a long time after Galileo. But a moratorium now would not be practical unless it were undertaken on a worldwide basis, and on this basis I do not think people would stand for it. Populations are increasing, expectations are rising, and there are many countries that are competitive and vigorous; science is felt to be a good and powerful medicine all over the world. For this reason, many countries are pouring their resources into the development of scientific and technological personnel and facilities, and the United States must inevitably be drawn into competition with them. So, of course, a moratorium, though possibly effective in the short run, is not likely to be acceptable to any modern or developing nation. Therefore, we must learn to cope with the problems science will bring.

One way we have already evolved to deal with such problems at the highest level of decision and control in this country is to provide the President with disinterested and effective scientific advice. This is a procedure whose development I can illustrate out of my experience in the last twenty years. At the end of World War II, President Truman thought he should have some sort of scientific advice, and set up a committee that met with him from time to time; I was on it, and I stayed on it after President Eisenhower took over. When President Eisenhower announced that we were going to launch a satellite, the Vanguard, we on the committee were very much alarmed, because we felt that the President and his lay advisers did not realize the risk they were taking with U.S. prestige. We tried to convey to him that this was a very difficult and expensive task, and that once a great power like the United States threw down the gauntlet with such an announcement, it could not afford to fail. But in spite of the President's announcement and commitment, the Department of Defense, which was opposed to the Vanguard, allocated a relatively small amount to the program.

When Sputnik burst upon the scene in 1957, the President realized he was in trouble. He saw that most of the people around him had not understood the effect the Russian success and American failure would have on the whole world. At that time I was the chairman of the President's Science Advisory Committee. Eisenhower called me in and asked for my advice. I said to him, "Mr. President, what I would suggest is that you appoint a man whom you like, whom you can talk to easily and whom you can be intimate with, to be your scientific adviser on a full-time basis. There's practically no decision which you have to make that doesn't have a very important scientific or technological aspect. You may not recognize it, but you need a scientist to help you in this. Almost every question you have to deal with has dimensions of science or technology."

The President then appointed Dr. Killian as special assistant for science and technology. (His successors have been Dr. Kistiakowsky, Dr. Wiesner, and more recently, Dr. Hornig.) President Eisenhower soon realized what a tremendous help it was to have such an assistant, because his other departments— Defense, Interior, Atomic Energy—did not tell him everything he needed to know about the scientific implications of major decisions. At last he had somebody who was loyal to him and helped him; very often, in fact, he felt he was all alone except for his scientific adviser. The role of this adviser has become more and more important over the years. Moreover, the function of scientific advice to the government in general has continued to grow. At the present time there is a Federal Council of Science with a representative from each of the different science federations, and there is an Office of Science and Technology, which has been established by law to coordinate and gather information. Eventually it was to become very clear to President Eisenhower and to subsequent presidents as well as to heads of cabinet departments and other agencies that scientific advice is something they need to make up for the basic educational deficiencies of their staffs. Whether this elaborate advisory system is a solution to the problem or just something we do until we find a better educational and administrative procedure, I do not know.

Another most disturbing problem is the inequitable distribution of scientific and technological progress over the world.

For a number of years we have been worried about the tremendous and growing gap between the developed and the underdeveloped nations. More recently, we in America have become concerned also with the relative imbalances in technological capacity among the developed nations—the so-called "technological gap."

Here again, perhaps, I can outline a possible solution based on work that some of us have been doing in NATO. It is of course simplistic to divide the world into the developed and underdeveloped areas. There is, rather, a kind of technological pecking order. For example, the productivity of the British worker is from 50 to 60 percent that of the American worker. There seems to be something basic in European society which produces this technological gap. In general, the European worker has only 50 to 60 percent of productivity of the American worker.

What accounts for this difference? I do not think the Europeans would be willing to admit that American science or American scientists are superior. And yet it is exactly in the application of science and technology that there is this enormous technological gap. How did it happen that the United States, which imported so much from Europe, has now, in the space of two generations, or less, become a leader? On the part of the Europeans there is a very deep self-examination and questioning about this gap. They ask, "What's wrong? How have we failed to use science and technology of which we were the essential inventors? How did it come about that we don't know how to use them well enough to realize the economic and social benefits that obtain in the United States?" I doubt whether we Americans could say that we have superior brains, or a superior society. But we certainly have a different society. We have different cultural traits; we take pleasure in exploiting inventions and new technologies, and we are good at it.

At any rate we have been deeply concerned with this problem. The political consequences are certain to become enormous if it persists. It provides the basis, for example, of a very powerful anti-American feeling. The inference drawn by many Europeans is that Americans must be very materialistic to be able to handle materials so much more readily.

Let me tell you, in this connection, about some small attempts to redress the imbalance. I happen to be the American member of the NATO Science Committee. This committee has one member from every country. We sit around the table; there is simultaneous translation; any vote that is taken has to be unanimous before it can become effective. It is a curious combination of science and politics. We have been concerned not so much with the military side of NATO, but with NATO as a cultural organization of member states. And to promote a greater cultural interchange in the field of science and technology, we spend about $2 million a year in fellowships for people to go from one country to another; we spend about $1 million a year in the support of about fifty summer schools in high-level science, which can go all the way from anatomy to zoology and to other more socially significant activities like operations research and demography. We also support research that is often done jointly by people from two or three NATO countries. Such combinations are part of an effort to weld these nations into a more homogeneous group. It is our thought that if two people cooperate they can do more than twice as much as two people can do individually; we call this the n^2 factor.

There has been, I think, a great increase in the effectiveness of European science in the postwar period. To a large degree, this has resulted from these summer schools and cooperative research projects to which Americans have contributed their technological knowledge and methods. We learned first from Europe, and now we are giving something back to our teachers. The Europeans are beginning to become quite independent in this respect, too.

But lately, the NATO Science Committee has evolved another way to close the technological gap. In October, 1966, we considered the formation of an institute of computer science. This is a very delicate political problem. The most advanced computers and the best computer companies are American, and we have stringent laws about the export of these machines because we are afraid that other countries will be helped by them to make atomic bombs. But there is the nonmaterial side of the computers, the software. Our committee has been thinking about an institute which would be devoted to perfecting the use—that is, the programming—of computers in ways to promote international cooperation. A single computer can be located in a central place with people to take care of it, and consoles can be hooked into it by telephone and radio, so that the result is a communal effort without the necessity of continuous physical gatherings of the community. This would constitute a very great advance, and has made cooperation more interesting to the countries concerned.

We have proposed this kind of cooperation before, or institutions that could achieve it, and almost always failed because of the rivalry between countries. But now, as we propose a European institution of this sort, the atmosphere is different. The very fact that there are technological gaps is forcing a kind of cooperation which has been almost impossible to achieve by argument or political means. The fact of advanced technology and its requirement that large units of people cooperate will force these national units to come together, I think, as they did in the Common Market.

I shall mention another idea being considered right now which is, perhaps, closer to the interests of many of the people here. Some years ago it was proposed, through NATO again, that there be created an Atlantic institute of science and technology which would provide educational and research resources strictly for Europe. This matter was considered by an international committee, and agreed upon, but national governmental officials in the end turned it down. It was turned down for simple, obvious, human reasons. No country could contemplate, with equanimity, the location of this sort of institution on the soil of another country. It was a negative competition. But our whole idea was to have an institution that would strengthen the whole of Europe, that would give degrees recognized in the different countries so that the graduates could move freely from one to another.

We are now trying to revive this idea in a much more favorable atmosphere. We have given up the governments as hopeless, and are trying to initiate it through private individuals with international concerns, through foundations, and so on. However, even the governments are growing more hospitable now, because they recognize the technological gap.

But as we contemplate hopefully the growth of international cooperation we remain uneasily aware of the great unsolved problem of our scientific and technological age: the growth of nuclear weapons. What worries me is that if I project our own American attitudes ahead a few years, I see only a small chance of avoiding nuclear war. If we keep on arming at the present rate, piling up all these weapons, it just seems to me against human nature, as I understand it, that they will not be used somehow. I am afraid some people think it is wasteful to make bombs and not use them. For example, prominent politicians, both Democratic and Republican, have suggested their use in defoliation and so on. The trouble is that when we get involved in a conflict and we have a weapon which some people think would settle it, it requires a form of self-denial not to use it. How long our restraint will survive under political pressure, ignorant as it may be, I do not know. I can only confess my growing anxiety.

Another danger comes, paradoxically, from the spread of the peaceful use of atomic energy. Many countries are desperately in need of power for industry and agriculture, and they are trying to utilize atomic power for these purposes. They will succeed, too, because they have to. The trouble is, though, that any sizable nuclear energy installation—even a peaceful one— will produce enough material for quite a few bombs in the course of a few years. It is only a question, though of course not a trivial one, of extracting the material, purifying it, handling it, and then forming it into the proper configuration. Enough is known now, one way or another, so that if the material is on hand, in time a nuclear weapon can be made.

It is certain, therefore, that many countries will have control of such material, and of large amounts of it. They may not have the most elegant bombs, as we have, but a primitive bomb like the one that was dropped on Nagasaki makes quite a disturbance. The critical point will be, I'm afraid, when nuclear power becomes available all over the world, and I think this point will be reached in about five years. We have, therefore, as I see it, only about five years to set the world's house in order on the question of nuclear war.

Perhaps the united Europe we are working toward will help toward this end. Certainly, it will help the United States and Russia to maintain their restraint on the use of nuclear weapons if there is a third powerful force in the world.

But even if we put our house in order soon enough to survive, we will still have to learn how to produce and maintain a balanced and humane environment. My prescription for this task is, as I said earlier, an educational one. It is my conviction that if we had a culture that permitted us really to understand our environment according to the deepest modern insights of science, we might be able to achieve the balance and harmony that will bring true civilization. A society cannot safely separate science from every other human value.

Naturally, I say there should be much more science in our education. But I do not think it will help at all if we introduce science to young people in the traditional way. Under this system, faced with a science requirement, the student takes the science of his choice: biology, anthropology, sociology, or, if he is tough-minded, some physics. On that level they are all equal. But what happens afterward? He will finish his science requirement and he may even get top grades, but, time being the great healer, his contact with science will soon vanish. The same kind of thing happens after the science student fulfills his humanities requirement. We need, therefore, a much more sophisticated integration of education, both for the scientist and the nonscientist. The study of science by itself seems to me quite limiting for many individuals. The pressures to specialize in a narrow field may turn the student away from the world. The same thing can happen in the study of literature.

We need, therefore, courses that relate science to the basic philosophical, ethical, and esthetic choices of human life. These courses would deal with such questions as how man's sense that he has free will can be reconciled with the dynamics of the physical universe. They would deal with the responsibility of the architect and the engineer to think of the human consequences of their designs and systems.

The aim of such education must be to prepare the graduate for the world in which he will live, to be able to cope with his environment, and to contribute to the strengthening and to the enrichment of his society. Strangely enough, the sciences, which are basically the motive forces for the vast changes we observe almost every year, occupy an ever-diminishing place in the program of the liberal arts college. In fact, a liberal arts education can with some justification often be characterized as an attempt at education in everything but science.

This paradoxical situation should be very seriously examined by those responsible for undergraduate education. It is not difficult to see why this situation has come about. The "battle between the two cultures" has been going on for almost a thousand years, and perhaps even before that in the remote mists of antiquity. The function of institutions of higher learning in the past has been more to conserve what is known rather than to contribute new knowledge. The function of conserving and passing on the knowledge of previous generations to the young is perhaps not very consistent with the function of discovering new knowledge. The poet, the painter, the writer, the composer, the law-giver have rarely been found on the faculties of institutions of higher learning, which have been more the place of the critic, the scholar, and the theologian. The essence of the scientific spirit is to use the past only as a springboard to the future. Good science is with almost miniscule exception current science, and the better science is the science of tomorrow. This cultural attitude on the part of the scientist, with all its vigor and arrogance, is therefore in considerable measure out of tune with the attitudes and concerns of the other culture. Conflict and dissonance is therefore inevitable. If peace and quiet should be the first desideratum in a college community, the gradual diminution of science teaching is to be hoped for. On the other hand, if the college community is to become a vigorous expression of the needs and opportunities of the age, the scientific culture is just the element which is required to provide the turbulence and creative energy needed for this purpose.

I would promote the study of science to overcome a certain kind of unsophisticated superstition. This superstition is manifested in our delight in moral absolutes. Except in science, we think in simplistic terms of one-to-one causation. So much of our thought is either black or white. Is it moral or is it immoral? Is it wrong or right? But actually, in real life we encounter a whole spectrum of possibilities and probabilities. It is therefore essential for us to have some feeling for the elementary mathematical notion of probability to help us make valid judgments. We need more quantified thinking in fields outside of business and science itself.

My prescription would include basic training in the language of science—mathematics. True, a notion of mathematical probability could have much to contribute to the exercise of choice among social alternatives; but more important, without a grounding in mathematics, much of modern science, in all its beauty, its relevance to human life, and its insights into the nature of the universe, is a closed book.

What I would wish for education is, I suppose, that it should bring about a fuller integration of modern society. The most important consequence of that integration would be, to my mind, a sense that we could exercise some degree of intelligent control over our environment. I would hope that we might lose the feeling I referred to earlier that change is just something that happens to us, that our scientific and technological advances are moving in unpredictable directions. I would hope that we could grow away from the feeling that humanity, culture, and human values are entities independent of, and bound to be overwhelmed by, technology.

There are great similarities, as C. P. Snow has pointed out, between our age and the Industrial Revolution. It, too, went on in an unregulated way; people were exploited in it without adequate return to society; society was bitter toward the machines. Artists, architects, painters, poets, and politicians of that age did not pay sufficient attention to the good life which technological progress should have brought about, and ignored the exploitation that it actually brought. The industrial revolution in England seems to me to provide one of the greatest examples we have of the failure of a brilliant society to integrate itself and to take appropriate action to remedy social ills. Indeed it probably could have been even more productive than it was if it had been more sensitively, more ethically, and more rationally conducted. With the benefit of that example, we in the twentieth century in the United States should be able to do better. Unfortunately, I tend more and more to think that the educated public, outside of science, displays a greater cultural lag now than it has displayed since, let us say, the eighteenth century. Most of the educated lay public are not as informed as they once were about science.

However, I do not advocate scientific understanding merely as a remedy for social ills. Such an understanding is an essential step to our finding a home for ourselves in the universe. Through understanding the universe, we become at home in it. In a certain sense, we have made this universe out of human concepts and human discoveries. It ceases to be a lonely place, because we can to some extent actually navigate in it.

I will confess that I have a slightly mystical feeling about this. We face a philosophical enigma: through mankind, matter has become conscious of itself. Think of it—*matter becoming conscious of itself*. Somehow, to me this is a very moving notion. It gives me a certain guarantee of the sanctity of human life. There, in the universe of matter, in the midst of it, is mankind, studying matter, studying itself, trying to understand itself. It is a noble endeavor. It seems, to me at least, to be an expression of man's essential nature, not all of it, perhaps, but the best of it.

OPEN DISCUSSION

QUESTION. You rightly ascribe much of the growth of American scientific and technological capacity to the support of research by government. But as a businessman, I would ask whether these funds are being managed to best advantage. Are not the uses to which they are put determined by the interests of the scientific community in the universities? This question is really a charge that the universities in the broad sense have not managed, to the greatest common good, those research funds that we, the taxpayers, have channeled to them through the government.

RABI. I think that is a very serious charge. I don't think you have anything to back it up with. To back it up, you would have to discuss the basic nature of our society, where we are going, and where we stand in the world.

In effect, when you invest in research, it's a long-term investment. And that's all it is—a long-term investment. You know perfectly well if you don't make it, you'll be out of luck. You've had enough experience in the past. And long-term investment in research is like breeding fish. There has to be a lot of eggs, and some of them may turn out to be fish, and others will just be eggs, or be devoured. This is the nature of the thing.

So you can start complaining. But it would be like taking a group of artists, and giving them work, a certain number of hours a day, and hoping they'll turn out good pictures. Just like the artist, the researcher has to be left to his own normal devices because you can't be like a Congressman who says, "Well, we'll give you this money—what will you discover?"

But there are two sides to this kind of support. One is the actual practical benefit which you may expect to get. On the one hand, you are buying growth stocks. On the other hand, and I think this is more important because we live not by bread alone, is the meaning of the research you are supporting. It is an expression of the vigor of the United States, an expression of the human spirit of adventure, the desire to know more. It's quite a new experience to see real money expended on research. And how great the expansion is you may not even know. As I have pointed out before, our department at Columbia, before the war, spent $15,000 a year on research. We now spend over $4 million. It's all government money.

COMMENT. I am glad you used the "we," the collective "we." It is government money.

RABI. Certainly, government money. And we ought to be glad for the opportunity, that somehow or other we've become smart enough to provide it, that we have public servants who do this. Because people didn't understand before. You had to look to Germany. Americans thought you could only learn about physics from someone by the name of Haasenpfeffer, or something of the sort. You've now got a scientifically independent country—a leader—and you know you gloat over it, you like it, you love it.

COMMENT. I don't like the $20 billion.

RABI. Well, you don't like the $20 billion, and you don't like to pay all the costs. You'd rather buy it wholesale; you'd rather get it some other way.

COMMENT. I want to buy at a discount, if management can save money. I have no idea of the direction this research is taking, in relation to the sector of the economy that I find all important. We are going to be a trillion-dollar economy in the early 1970s. This means that in the sector of consumption, there is going to have to be $300 billion of new product consumption. Where are these new products coming from? They are coming from industry, to a degree.

RABI. But what is industry? Industry is people who got educated at universities and who employ people from universities. There is nothing special about industry. It is simply a group of managers of other people's money!

COMMENT. So are you! So are you!

RABI. You are not a different species. You are just people like us. It reminds me of my good friend Dr. X (a physicist). Every time his bomb exploded, he took a bow, but the bomb came out of technology, and some people, including Dr. X, were clever enough to get in on it. They bowed as if they somehow did it, when they were really only riding the train. You can read biographies of very perceptive rich men who know that that's all they were doing when they "struck it rich." They happened to be in the subway crush, in the right place, when the train came along and somebody pushed them in. They know that perfectly well. I can give many examples, and so can you, even more. So let's not make this distinction between industry and the others.

QUESTION. You said that the educated layman of today still entertains a Newtonian view of the universe, while modern physical science has been based on principles of relativity and indeterminacy. Then you gave an educational prescription that would correct this anachronism. As a layman, I would like to hear a little more about how a modern physicist sees the universe, the world of matter, and man's position in it.

RABI. The universe is man's environment; we've really just discovered this. It is not just the planet, or the surface of the planet, it is the universe—whatever that term may be—the totality of things, which we are trying to understand, which we physicists are trying to fit, somehow, into a scheme. We are trying to find out how things really work, why things are as they are, what are the basic regularities, if any, which govern this remarkable structure of which we see just a very small part. We want to know what is it that makes the universe, and what makes it persist. We know of many physical structures that live for a millionth of a second, a billionth of a second, a trillionth of a second, and, in some cases, a ten-to-the-minus-twentieth of a second and even less, structures, which during that period, are very solid, and then come apart. But here we are faced with this remarkable thing, the universe that we perceive with our unaided senses, which had seemed so stable, so enduring.

Our concern with the environment, therefore, must be much more basic and thoroughgoing. When we conceive of matter, we must conceive of matter with respect. Because matter is not mere matter, and matter is perhaps not even tangible except in a very limited sense. It used to be believed that the atom was, as indicated by the etymology of its name, indivisible. But this limitation is only semantic. Once, in Germany, I think, when Professor Sommervell was discussing the structure of the atom, a philosopher got up and explained that the atom was really indivisible, that you can't talk of its structure; so Sommervell said, "I agree with you. Let's call it Tom instead of atom."

The atom, one discovers, is empty. Only a very small fraction of its space can be considered as occupied by its constituents, that is to say, the electrons and the nuclei. And further investigation, with new methods of high-energy physics, has established that the electrons and the nuclei themselves are in no sense hard, impenetrable objects. They, in turn, are composed of much smaller units, which, in turn, as one investigates them, turn out to be divisible. Thus any semblance of a boundary below which matter is indivisible, or of some solidity in the kinesthetic sense with which we've grown up—that boundary disappears. How we will ever come to understand all this, I do not know. We are very, very far away from understanding it. Is matter actually some state of space or some convolution of space? Whatever it may be, this world is a much more mysterious thing than Horatio contemplated in his philosophy.

The properties of matter are seen differently on every level. On the ordinary, biological level on which we live, matter is hard and impenetrable and solid and so on. You go further down, and this solidity begins to disappear more and more. We've not begun to plumb the depths of this. A whole new world has been opened in the last quarter of a century, in which we've talked not only of the structure of the atom but of the structure of the units of the atom.

QUESTION. If we do succeed in controlling nuclear weapons and using science and technology to create a materially comfortable environment, what can we do to solve the problem that you hinted at, namely, having a meaningful life for great numbers of people with lots of leisure time? This may be the great educational problem of our time.

RABI. You ask what really makes life meaningful, and it comes down to a basic question. I think you have to look at people whose lives are meaningful, and you find that they are mostly living lives which take them out of themselves and get them interested in things. The most meaningful sort of life about which we have written historical experience is the religious life—the worship or contemplation of some other being, something larger than ourselves. Then there is patriotism, the community, or doing good for people. All these things can become meaningful. But when you produce the ultimate of the good material life, you may have lost many of these activities; they may be irrelevant. A religion which would stand the impact of science really may come, but I doubt whether it is here. But there is one thing which I think could make life very meaningful as a community effort. We see one small example of it, and a somewhat vulgar one from my point of view, in the effort to explore outer space. Imagine, billions of dollars going into that project and the American people getting an enormous kick out of it. If they didn't, Congress wouldn't appropriate the money needed. There are enormous sums, more than all the research in all the universities, going into this particular program. There are profound meanings and desires, of which I see this as an expression.

When I look at the future, the utopia you are talking about, there will always remain an endless frontier of human effort and human experience. And I'd like to see the general educated public have a part in it. At this moment they're disengaged from this big effort. They pay for it, but they have no real part in it, no real understanding and appreciation. And nobody really tries to bring them in very seriously because the kind of education which would bring them in is lacking. We're just filled with the old idea that the population consists of an elite and a mass. We begin to feel, "Maybe the masses can't understand it all." That I don't believe is true. So I think there will remain this endless frontier of science, in which we try to bring understanding to everyone. Now, what we can do with the smaller proportion of people who may not have the actual capacity to understand, I don't know, but our problems are far from that point yet. But we may come to a world in which there are two classes of people—I don't know.

QUESTION. If the advance of science makes it difficult to have a religion that is relevant to modern life, what becomes of our need for values? Can science perhaps show us some absolute on which our moral, ethical, esthetic values can be based, something other than tradition or habit?

RABI. Science and technology cannot be separated from every other human value, and there may well be lessons in physical science from which human values can be derived. For instance, take my distinction between the Newtonian and the relativistic world pictures. Professor Berlin has already outlined vividly the consequences of deriving social values from a too-confident acceptance of simple ideas of causation in the physical world. I can reinforce his lesson from the insights of relativity. If I study a subatomic particle, I will want to know two things about it: its position at a certain time, and its momentum; momentum is merely speed multiplied by mass. But if I want to locate its position accurately, I find that in the very act of doing so that I'm influencing it; I'm bound to give it a momentum unknown to me, indeed unknowable. Or I can give it a known momentum and the more accurately I know the momentum, the less accurately will I know the position.

The two things that are required to measure the position and the momentum cannot be ascertained at the same time. You can have one or you can have the other, but you cannot have both. One set of values can be entirely contradictory to another set of values. There may be one set of values which in themselves are not contradictory to one another. This complementarity is like that mentioned by Professor Berlin in the case of justice and mercy. You can't have both in this sense. I'm not making a plea for science as a substitute for religion here but to show that these basic ideas are reflected in almost every human effort. In physics they can be made precise. That doesn't mean we understand the mystery behind it—just what is the nature of particles that does this. It is very interesting that Einstein himself, who laid the foundations for this concept, never cared for it. He always felt that it was something provisional, something you do until the doctor comes, so to speak. But we have not been able to get away from it. And, it does have a profound similarity and perhaps a causal relation to what we know of values in human affairs, after all, what are we except a collection of matter of different forms?

Take the principle of indeterminacy a step further. Because I cannot make simultaneous measurement on this scale which will permit me to predict with certainty the position of a single particle, I must study motions of many particles. I will end up with frequency curves showing how this kind of particle is likely to act. I can then predict what will happen in a general way; but what will happen is not any one thing. There's a whole series of possibilities, just as Professor Berlin said happened, when you work with a whole social system with a particular goal in view. You don't usually reach it. You could reach it. It's not impossible, but most likely you will end up with an unexpected result. And this is the lesson that we have from physics, which I think we must absorb very deeply in our social thinking; that when we deal with a social system, the outcome is variable. We might, by knowing a great deal about such a system, predict a whole series of possibilities. And we will do a great deal better when we are able to assign probabilities to those possibilities. In essence, though, there is always this necessary unpredictability.

As for the future, although I cannot predict their exact nature, I think science may give man new values. I think our values change with our knowledge. If we can control animal or human evolution, there will undoubtedly be in the vast variety of human beings those that will choose to control it. The problem, I think, may become a very serious one. I've often wondered what would happen if some superbeing arose out of such control, either out of the human race or out of the animal kingdom. We'd probably kill it, perhaps for so-called moral reasons. At any rate, our interests may turn, at some future time, to concentrate our efforts on maximum knowledge and maximum understanding. There are hints of it right now, in the vast amounts of money we are pouring into space exploration. Some people say it's a contest with the Russians. It's partly that, but it doesn't quite explain the tremendous interest with which Congress, which is often cold to purely scientific interests, pours out the money. So there is a real, deep interest which one could not have foreseen twenty-five years ago. Our interests and our values have changed. When such a thing happens, when possible, I think, there will be a very great interest in what the direction will be. It will be decided by the consensus of either a large or small group.

About making use of our discoveries, about their application, I think this is a very serious matter. I think there is a real lack of a kind of forum or organization where utilization could be discussed. I am not one of those who believes that every scientific invention has to be exploited to the full, even if it will cause tremendous inconvenience, or death, or embarrassment, or humiliation. I do not think it is necessary. We do not have to apply everything that we know, just as we do not have to do everything we can do. If we have a gun, we do not have to go out and shoot half a dozen people, though we could. About these things, there ought to be some way of providing provisional guidelines. And these should correspond to the particular morality or custom of the time.

I suppose that moral life revolves around choice, and the right choice cannot be made without knowledge. The mother can love her child and try to do her best, but simply poison the poor little beast by not understanding what's happening. It's happened again and again. And so on, through life. By and large, our knowledge comes to us through action, whether physical, psychological, or otherwise. Knowledge gives great scope to moral choice. So I think the scientific contribution is really basic. Think of all those poor old women who were tortured for witchcraft in Salem. The people who did it meant well. There are many similar examples. So I think that as we increase our knowledge, we raise our morality. Our morality at the present time, in spite of all that's going on, is very much higher than at any time in the past.

Always, though, I come back to the kind of vision that science can give man, which can help to reconcile him to the universe which is no longer so easily explained as it once was. We certainly have learned a lot about the world in the past few centuries. We recognize how profoundly subtle and mysterious it really is, how different everything is from the ideas we have from immediate experience—not only how vast it is, in its dimension, but in the other dimensions, so to speak, of subtlety, or originality of nature. We stand in awe of all this and we stand in awe of the human mind that could make these discoveries, that can put these things together, that can discover notions that are so very different from traditional ideas and structures.

And, through understanding the universe, we are at home in it. In a certain sense we have made it, out of human concepts and discoveries. It is not a lonely place, in the sense that we can actually navigate in it. What has happened is that we can be at home in the universe. It's our universe. And that's exactly why I am a scientist.

DISCUSSION: I. I. RABI AND JACOB BRONOWSKI ON NUCLEAR WAR

QUESTION. (put to Rabi and Bronowski jointly). I would not necessarily agree with Professor Rabi that the proliferation of weapons will inevitably lead to their use. I think we're reaching a point where their use is less likely than heretofore. At the moment, atomic weapons are in the hands of five powers, and the mental health of the two most powerful and dangerous of the five seems to be improving rather than getting worse. Would you gentlemen comment further on Professor Rabi's prediction in the light of that observation?

RABI. I'm glad you used the work "likely," because I didn't take an absolute position. I said that if one projects from the present, knowing how human beings react, one would have to be pessimistic. We have an enormous stockpile of weapons. We have a number that is in the tens of thousands, any one of which could take out one of our big cities. Well, if we have tens of thousands of these weapons, I presume others will, too.

So we are living in a world which is on a hair trigger. And as time goes on, the failure of these mutual deterrents becomes worse and worse. In other words, at the present time, the safety of the United States is very much less, after we've spent these hundreds of billions for our defense, and will be very much less in the future.

I cannot share the optimism of those people who say, "After all, this is so serious that people won't do it." Such a statement is based on a belief in rational action which is belied, every day, by the countries and the statesmen involved. All we have to do is listen to the speeches in the United Nations, or in our own Congress, or out in the street. I personally feel that the safety of my country should not depend on so light an assurance as that man will act rationally.

You remember Pascal's dictum that, on the chance that there might be an afterlife, he wouldn't take a chance of eternal damnation for just a short lifetime of convenience in not following the rules of his religion. I would say that our stakes in the event of nuclear war are so high that we should expose ourselves to the inconvenience of positive international action to achieve our salvation.

What concerns me is not that all life would be wiped out, or anything as grandiose as that. I'm thinking specifically about our own United States. If you really love the country, the features of its physical existence and these remarkable, wonderful, peculiar institutions that have arisen in it and which were started by a unique set of people, the founding fathers—then think what would happen to this, this creation, this going concern. You can't stand the feeling that it's in such danger. I'm not so sure that there is only a slight likelihood that somebody will do something mad, especially when I remember, only a short time ago, the real confrontation over Cuba. Once that started, and President Kennedy had made his statement, I stopped reading the paper. The course of events was set; he couldn't back down. We were that close to it. Then Mr. Khrushchev backed down. If the shoe had been on the other foot, would we have backed down? I doubt that. I really doubt it. I say this just from my own estimate of the nature and caliber of the people. That's how close it was. And this situation can arise anytime, anywhere in the world. This is without proliferation, just the Russians and ourselves.

Now, this situation becomes complicated. Many other countries come in. Perhaps their boiling points are lower, and they are more likely to use weapons of this sort if they possess them. These weapons become more usual. Some small country might even threaten the United States. When you begin to look at that, your blood runs cold, because you begin to doubt the value of all our endeavors—our system of justice, the Supreme Court, the President, our whole system of morality, everything. I doubt that these would survive very long, under the impact of an atomic war. The result of survival will be each man for himself, the disruption of our whole national organization.

It doesn't take very much imagination to see what would happen. Only a few years ago, there was a hurricane, and some bridges between New England and New York were wiped out. How much destruction, and how long it took to repair that! After all, what happened in the blackout in 1965? Some switches went wrong and the whole system broke down. You couldn't produce, you couldn't work. Now just picture the integrated system under which we live. We have an incredibly complicated organization. Just cut it at a few points, and it's gone. I'm not even talking about the destruction of human life, but of the destruction of institutions. Destruction of human life would follow.

I'm not content to sit back and say, "I don't think it's likely." I want solid, solid organizational things to deter it. And I do not understand how our Congress can fail to worry about it in the same way as I do, and make this a very high-priority affair. I do not know what it is that makes Congress feel so cool when I'm so hot on the subject. If you're keeping your head when all around you are excited, maybe you haven't understood the situation.

Now, why did I say that the safety of our country is less than it was before we spent billions of dollars for these thousands of weapons, any one of which could destroy any one of our great cities? It is the result of the arms race we've had, which has created so many nuclear weapons on both sides. If we'd had a nuclear war, say, in 1950, when we had about fifty weapons, there would have been tremendous destruction, and people who lived through it then would have a vivid image of how bad it could be. But when you begin to compare that with the devastation from tens of thousands of such weapons, with people talking of antiballistic missile weapons that, while stopping an incoming missile, would nevertheless probably flatten the whole city being defended—when you begin to talk that sort of game, and, really, intelligent-sounding people with high salaries and well-tailored uniforms are talking about these things —you begin to wonder what sort of world you are living in, anyway. What sort of madness has taken over? Not only on our side, but on the other side. Why are these negotiations going on interminably? One is not willing to give, here or there. One is not willing to take a slight chance either way, for the larger opportunity. This is what I meant. That is what makes a break-down more possible.

I would like to add one little point, just a personal judgment. Suppose, under certain circumstances, we did use nuclear weapons, let us say against China, quite safely, and there was no retaliation. But suppose that we did an enormous amount of damage.

How, I ask myself, in a country like ours, built up as it was on basic principles such as are in the Declaration, the Preamble to the Constitution, principles of justice, humanity, and the like —how would we feel afterward? Wouldn't such a successful war basically change the character of the American people? I ask myself this question. What if we got away with it? Hiro-shima gave us a little taste of it, but after all, we were at war, we were first attacked in a cowardly, unjust way, and there was a history of atrocities, at one time, to our soldiers in the Philip-pines. All these facts built up a human relationship which very much diluted our sense of guilt.

But how would we feel after the hypothetical devastation of China? In a certain sense, I'm afraid—and this is just a per-sonal reaction—it would be the end of the American dream, in the sense that we are a nation based on justice and right rather than on a master race whose will must prevail. I have a feeling that is very disturbing, that the very existence of this enormous power in our own hands calls for an unlimited amount of re-straint and reserve. We've shown it, it is true. The time when MacArthur was so badly hit in Korea, we could have gone out and bombed the Chinese side, and wiped out a few of the cities which were training bases for those troops. The fact that we didn't, I think, is really great. We could have, in view of the provocation. How long such a spirit will last, I do not know. I think our patience is wearing thin. It is very difficult for a strong man with a big club to hold himself in after a lot of frustrations and irritations.

But even if the nuclear powers control themselves and each other, what about the other states? Take, for example, the Indians. They say to us, "You—meaning the United States and Russia—you want us not to make atomic bombs. But you're not giving us any kind of guarantee for our safety. And furthermore, why should you take an attitude of superiority over us, that you can possess them and we cannot? If you take this position, that we should not make atomic bombs, you certainly have to take measures to reduce your own atomic armaments. Otherwise you are saying that we should be a second-rate power. We should not have any choice and must put our fate entirely in your hands."

This is a very strong argument.

And France has gone ahead to act on this argument. France has simply said, "The hell with you. We know how to make them. We'll be responsible for our own safety." Back in 1954 I spoke to a Frenchman who is very high on the French Atomic Energy Commission. He said, "If a test-ban treaty is signed before Christmas, we will sign it. If it's later than then, we won't sign it until we've tested our first atomic bomb. Then we would sign it." Just before that, France had decided that this was the sort of thing they had to do, rightly or wrongly.

In other words, I have a feeling that we—that is, the nuclear powers—have failed to give the moral leadership which was required. I think we in the United States felt very good, and very clean, morally, because we haven't used this against anybody after Japan, and that somehow or other, we were entitled to this unique position because of our moral nature, our culture, and our institutions.

BRONOWSKI. I wanted to evoke this elaboration of the discussion, from Dr. Rabi in particular, because I think these things are not sufficiently thrashed out. We are all very frightened of these things, and people prefer not to talk about them. You see, both he and I are the only people in this room who have ever seen any of these weapons used, and both he and I came back with an enormous sense of revulsion. I think it's very fair to ask us, "Well, what do you want to do?" I think none of us know what we would do, but I think we are all very conscious of the fact that, to make people feel that something ought to be done, the urgency of the data needs to be presented.

RABI. If you really want to give such measures a high priority, then you invent steps leading toward them. If people really want to come to an agreement, then they'll make one. I do not think there's any one gimmick. I, too, am talking about the urgency of the need. I do not think that any of you people in management, once you knew you wanted to do something, could not find a way of dealing with a potential enemy. And the prospects are better than usual here, where you have mutuality of interest. The Russians are just as scared of us as we are of them, maybe more so.

BRONOWSKI. The mutuality of interest is there, but it has to be recognized. Now, I remember William Lawrence, in 1945, in a conversation, saying to me about the atomic bomb, "This is such a weapon that human beings will just have to stop making war." And in some way, I still feel that if people would realize just how great the danger is, we could reach the situation where we could say, "No, none of us could use this in war. It ought to be destroyed." And then we would work out the steps, as he says. Starting with the objective, we would work out the steps.

Five nations now have working weapons which have actually been exploded. Certainly, as many other nations are working on weapons which they haven't yet tested. Professor Rabi said that we are now in greater danger in this country than ever before, and somebody questioned this, but I want to bring this home to you very clearly. You see, there was a time—let's make up a kind of joke about this—when, in a war between the Philippines and the mainland of the United States, there was no doubt who would win. Sometime in the next ten years or so, this balance will be disrupted. You see, there was a time when, if you had 10,000 times as many soldiers as the enemy, you were bound to win. But it is not true that if you have 10,000 times as many atomic weapons as the enemy, you are bound to win. The fact is that their one-tenth may be quite enough to strike what is essentially a mortal blow. And unfortunately, this makes the finger of a small nation itch on the trigger much more than that of the big nation. Remember Pearl Harbor. The Japanese were fanatic enough to think that they were going to destroy the United States by that surprise attack at Pearl Harbor. But just imagine what the effect of that attack would have been if both sides had had atomic bombs. The Japanese could have had one-tenth the atomic bombs of the United States, and Pearl Harbor would have been an infinitely bigger disaster than it was.

And this is the situation we have to face: a world of these nations. Perhaps you won't think it inappropriate of me to tell a small, ridiculous story that will lighten the atmosphere. I remember being in Israel one time, and hearing a story which went the rounds, of two Israelis who met in the street. One said to the other, "I've solved our economic problem." The other said, "How did you do that?" "It's simple. We go to war with the United States, we lose, and then they build us up." And the other man says, "But suppose we win?"

Well, that's a silly story. But one day soon, there will be a situation in which Israel or Egypt or the Philippines or some country of that sort can say, "Suppose we win?" without its being ridiculous. Of course, there would be no victory in this, but it could be a very small country that would ask. And it is this disequilibrium, which has been caused by atomic weapons, that the old generation, which grew up with ordinary weapons, hasn't really seen. This I think is why a sense of urgency about this thing, which Rabi happily advocated and which I've been anxious to underline, should inspire us all.

There can be no end to this except the abolition of atomic weapons. Of that, I'm sure. I'm also sure that you can't go out and say, "I've abolished atomic weapons. What are you going to do?" There has to be a procedure. But if you realize what the end must be, if great civilizations are to stay, then I think you have a different sense of urgency about the need to work for it.

The Impact of New Science

JACOB BRONOWSKI

Dr. RABI AND I are both addressing ourselves to the theme of the impact of new discoveries in science. My concentration on the new biology, however, has less to do with its being biology than with its impact on human society. First, I am going to say something about what is new in biology, what techniques will change our lives in the foreseeable future—by which I mean, roughly, in the next fifty years. Second, I am going to say something about the social consequences of these new techniques. And finally, I want to show that when we have discussed technology and its social impact, we have only scratched the surface of the profound way in which a new science alters the conceptual universe of man and the range of his ideas and his emotions.

First, then, where are we in biology and where are we about to go? In the last fifteen years, a great change has come over biological research. Biologists have become convinced that what we want to know is to be found in the individual cell. And they have become convinced that it is the physical structure of molecules in the cell that will tell us what we want to know.

The great revolution in physics began about a hundred years ago when Dmitri Mendelejeff predicted the appearance and behavior of new elements, and so opened the vision of modern physics by pinpointing the atom as the potential key to

JACOB BRONOWSKI IS SENIOR FELLOW AND TRUSTEE AT THE SALK INSTITUTE FOR BIOLOGICAL STUDIES.

understanding matter. In the same way, in the last fifteen years, we have grasped the biological cell as the basic unit from which we must explain all living actions. The part the atom has played over the last hundred years in directing the form of physics is parallel to the part which the cell is now playing as the object whose behavior we have to analyze—because its behavior underlies everything else in the body.

A biologist studying the cell says, "Well, I've got to show how one molecular arrangement becomes another and what influences it." And this approach applies to the study of genes, or of viruses, or of antibody resistance to disease.

I have said so far, by way of preliminary, that the double key for recent biological research has been the concentration on the cell, and the unlocking of the mechanism within the cell purely in terms of molecular structure. Indeed, most of my colleagues in biology now are ex-physicists, who came to biology with the notion that what can be done in physics can be done in biology, provided a simple enough unit to attack is found—a unit that underlies the whole animal. Let me repeat that: provided a simple enough unit to attack is found, the methods used in physics will do in biology as well. I must say biologists have scored a most remarkable triumph in the last fifteen years.

Let me, then, talk in a technical way about the things that have been done and the things we may expect to do. A single practical development that has transformed biological method and thought is our ability to conquer invasive diseases. From time immemorial, mankind has been fighting those diseases that are produced by outside invaders—by the attack of bacteria and of viruses. If I may be allowed to draw a slightly fanciful picture, we were all on an animal hunt, and the animal to be hunted down was the microbe that Louis Pasteur first recognized as an attacking invader.

But the fact of the matter is that, in that sense, medicine has completed its basic task. In that sense, the doctor is going to be reduced, within the next twenty or thirty years, to the status of a druggist. Just as we now go straight to the druggist to get something for indigestion, so we will no longer bother the doctor to deal with these invasive diseases—for a druggist or a nurse will be able to do that, almost mechanically. The diseases themselves are no longer on the forefront of knowledge. Research in this area is no longer the most exciting research. The invasive diseases may be regarded as defeated.

This may seem a heartless thing to say, when so many in Africa are suffering from insect-carried and water-borne and infectious diseases. But the fact is that we could eradicate these diseases in short order if we were willing to make a sufficiently large investment of men and money. In the West we have done this to a considerable extent. For example, when I was a boy, the two great killers in hospitals were pneumonia and gangrene. One of the reasons for not going to a hospital, in general, was that the best way to avoid these two killers was not to come in contact with them. Now we go to a hospital, and neither of those infections is a serious danger any more.

If this has happened in medicine, what are we fighting now? The whole aspect of biological research has changed to an analysis of those things that go wrong in the ordinary and natural growth, division, and aging of cells. If I may use one simple word, which I hope will be seen in its context, we are no longer fighting disease, we are fighting *age*. I do not mean that we are fighting the falling out of teeth, short sight, shortness of breath, and so on. I mean we are interested in discovering what happens to individual cells as they go through their life cycles.

To illustrate this change in a most graphic way: the Salk Institute owes its existence to the wonderful solution to the problem of polio, a comparatively rare and dramatic virus disease. But what are people interested in now? They are interested in heart disease and cancer.

Neither of these diseases, above all cancer, is a disease in the

sense in which polio is a disease. Heart disease is something that
happens to you; and cancer, so far as we know, is something that
happens to the cells in the normal course of their growing old
and being replaced by others.

Let me say a very simple word about cancer. Let us take a
very clearly differentiated and analyzable human organ, the
liver. The liver is an important chemical factory; and so it is not
surprising that in any mammal, if the liver is injured, that liver
will sustain and repair very extensive injury. But, if you lose a
finger, you must do without that finger. Nature has not invested
a large amount of biological capital in replacing the finger. Its
general attitude is that if you got on with ten fingers, you can
get on with nine. This is not true of the liver; and so, if you
suffer liver injury, a vast apparatus is marshaled at once to repair
it. You can take a liver, certainly in experimental animals, and (it
is supposed on good evidence) in human beings—a liver which
has been damaged to such an extent that only one-third of the
total liver survives. The body then sets about repairing and replacing the damage. Within a matter of ten to twenty days, the
whole of the liver has been reconstituted.

To do this, liver cells have to be created: the body has to
make new liver cells. So, almost incredibly, the existing liver
cells suddenly start to multiply. The body cannot start with
nothing—it cannot make liver cells out of nothing. But out of
the healthy liver cells that are left, a tremendous process of
multiplication begins, making new liver cells in the body. This is
unusual, because normal liver cells have a long lifetime and very
seldom need to divide.

During this process of repair, however, liver cells multiply
at roughly a thousand times their normal rate. And that is just
about the same rate at which cancer cells multiply—a thousand
times the rate of normal cells. So, in a way, the replication of
liver is a kind of cancerous process. The liver suddenly starts a
population explosion of cells. And if it were cancer it would go

on. But the really miraculous thing is that when the liver is back to its normal size, the growth process switches off. The job has been done.

The whole cancer problem would be solved overnight if we could make the cancer cells switch off. Most biologists think this is the crucial feature in the complicated processes of cancer—that the "off" switch has gone awry. The cell multiplication has lost the control mechanism which switches off when the need which set it going has passed.

Biological research in the next twenty, thirty, fifty years, in my opinion, will concentrate on this concept. We are trying to understand each process, to find its natural mode of control, and to apply the same kind of control to other body processes.

As you know, there have been grave criticisms of the work on chemotherapy as a treatment for cancer, on which the national agencies have spent a great deal of money. Doctors and chemists alike were beginning to say, "That's surely not the way to tackle cancer." I do not know whether they are right or wrong; nobody knows what will turn out to be right or wrong here. But it is significant of a profound change of approach to biological problems. People are now saying, "Let's stop hitting cancer over the head with a chemical club in the hope that it will lie down and die. Why don't we get inside it, and understand what makes an essentially normal human function, namely cell replication, go wrong, produce the wrong kinds of cells, and fail to switch off." Thus, our whole approach in biology for a long time to come will center on *active* health. The new outlook is that we should not be content to cure only what goes wrong when the body is invaded by some outside agency. Instead, we should ask how we can direct the ordinary body processes so that they do not go wrong, and so that little oddities which start here and there are redirected before they go wrong.

I would like to tell a slightly irreverent and irrelevant story here. When Isaac Newton had finished his great work and had shown that the universe runs like a clockwork, he was asked what part God had to play in the mechanism of the universe. Wouldn't it run down? Didn't God have to wind it up? "No," he said (more or less—I am making a free paraphrase of his thought), "God wound it up and now it's running fine. But," he said, "I just don't believe there's any natural system which doesn't from time to time need a small masterful touch to keep it tuned in working order." In a way, we have just that vision of what real biological control of human growth should be.

And that, of course, goes for the whole cellular mechanism. It isn't just a question of cancer cells, to which a rather dramatically exaggerated importance is attached. We shall certainly want to understand the mechanism of heredity in the same way. I may as well go into this now, because the minute I sit down, people will want to ask me whether we are going to grow children with three heads. This kind of biology is always feared, on the ground that some deviant species is going to be started. Sometimes I imagine that the first monkey couple to produce an extraordinarily bright child who looked like a step in the direction of becoming a human being asked each other, "What have we done?"

Let us take a characteristic phenomenon. Within the last twenty years, it has been realized that mongoloid children suffer from a simple but extraordinary malfunction. Ordinarily, a set of chromosomes from one parent and a set from the other parent mesh, and you have two chromosomes at each locus. For some reason, which we do not yet understand, at what is called locus 21 something unusual sometimes takes place: three different chromosomes appear at this locus, two from one parent and one from the other. When that happens, all kinds of things go wrong with the child, and it becomes, in appearance and behavior, a mongoloid child. There is nothing wrong with the child so far as his genetic outfit is concerned. All the rest of the chromosomes may be full of the right things for the right blood groups, for being a genius, or for being a baseball player—all the things one would like for a child. But in some unfortunate way, at locus 21, three chapters have been bound into the book instead of the specified two, and everything is then thrown out of gear. We have no idea how this happens. And so we have no idea how to correct or prevent it.

In the last year or so it has come to be suspected that there are other human diseases which originate in the malfunction of the chromosomes at locus 21. Leukemia and other forms of cancer appear to be more frequent in people who have something wrong at this locus 21. It is clear that we have put our finger on what you might call a hot spot in the assembly of 46 chromosomes. Whether we shall be able to do anything about it is not yet known. This kind of investigation, however, now dominates the minds of even quite traditional biologists. And it is a most exciting notion that one can see, as it were, in the book which contains the hereditary blueprint for our lives, that certain pages or certain chapters are particularly liable to go wrong, and that the study of inborn health or sickness would be particularly useful at these places.

I ought not to stop this part of my technical description without saying that there are a thousand and one other inner processes of the body that we shall learn to understand and improve. You already know about grafts; for example kidney grafts, which, although spectacular, have also been too often unsuccessful. Grafts are a real problem, a fascinating problem, on which I will spend just a little time.

The reason grafts are frequently unsuccessful at present is that the body has an extremely high personal identity; it is directed, from the outset, to resist the invasion of foreign tissue. When evolution, or nature, or whatever you like to call it, assembled the human body, it built into the body from the outset a system of recognition designed to keep out foreign protein. The notion that you are individual is built into your chemistry from the word "go," and you are not like anybody else. You really are not like *anybody* else. If two people in this room each gave me a square millimeter of tissue, a good biologist could tell them apart. And if he could not tell them apart, he would come back to me and say, "You know, I have a feeling that you have been cheating. Either you've given me two pieces of tissue from the same person, or you've given me two pieces from identical twins (who have been formed from one fertilized egg which has split in two)." The body has a high chemical specificity, and the purpose of this is to resist the incursion of other things. If you happen to suffer from an allergy which makes you break out in spots when you eat strawberries or eggs, this is an expression of the body chemistry which is designed to prevent incursion or invasion.

Allergy on the one hand, and the rejection of foreign grafts on the other, are both the work of an apparatus of protein recognition and resistance in our cells which causes them to make antibodies. We would all die of diseases if the cells did not recognize invading protein and build defenses against it. Yet when you have an allergy, it is not strictly the invading protein that upsets you. It is your body's effort to neutralize the invading protein. All resistance to disease is of this nature. If something is grafted on you from another person, your body reacts in the same way and tries to reject the graft. The most obvious example is that in a blood transfusion we have to be sure that the added blood is really close in character to that of the patient. Although blood chemistry is comparatively simple, there are still quite a number of chemical characters that we have to match and get right.

There are an enormous number of things that have to be right to get a successful skin graft. If your hand is burned, you might expect that a piece of skin will be taken from the hand of another person, put on your hand, and then watched. For ten days or so it seems to be doing fine, and then it curls up and falls

off. This is the extraordinary situation: that if you have a bad
burn on your hand or your face and want to mend it, the sur-
geon does not go to somebody else and take a piece of skin from
exactly the same place. No, he takes a piece of skin off your behind and puts it on your hand or your face. So far as the body is
concerned, your behind is you and is as much you as your face
and your hand; whereas somebody else's face or hand is not you,
and is rejected.

If in the future, we are going to make a usable technique of
grafting, we have somehow to find what the limits of this rejection are. As you know, it is possible to make blood transfusions
between different people, provided we get the chemical constituents right. So it will surely be possible, in ten or fifteen years'
time, to make quite extensive grafts provided there are certain
parts of the chemistry that we get right. The body will not insist
that we have to get all the billions of different chemical units
right. Probably if we get twenty, or thirty, or forty right, it will
accept it. At any rate, this is a very important problem to study.

I have given you this glimpse not to dazzle you with mar-
vels, but simply to make one central point: today there is a
different approach to the problems of health. And a tremendous
technology will flower from this. When, from time to time, the
British government would say to me, "What kind of technology
ought we to invest in? Ought we to start a great project on computers and so on?" I would say, "Computers? Computers? I
think that problem was solved in principle by Charles Babbage
in the nineteenth century. That is all classical. If you really want
to make a great investment in the industries of the future, you
had better get all the biologists and biochemists together to see
what will be important. It might be something trivial, like contact lenses, or it might be something profound, like antibody research and the testing of substances which allow transfusion." I
believe that there lies the technology of the future, as a seed in
the modern science of biology.

I would like now to turn to the second part of my subject. The social consequences of these changes are sure to be far-reaching, and it is no use putting our heads in the sand about this. Of course, we never know with certainty what the social consequences of any discovery will be. Who would have thought that the unfortunate character who invented photographic film would have been responsible for the California film industry? And thus indirectly for contracts which prevented film stars from having affairs that might give rise to gossip and scandal? That consequently stars led their love life in public, by repeated divorce and marriage? That therefore the beautiful pin-ups of films in time became the models of the divorce business? And the climax, that one-third of all marriages contracted this year in California are going to end in divorce—all because somebody invented the process of printing pictures on celluloid strip.

On the same lines (which I leave you to trace), who would have supposed that Henry Ford's devising of the sequential method of assembling a motor car would finally result in upsetting the whole moral code of the American middle classes? For it is evident now that the car provided young people with more privacy than the home, and that as a result it became usual to begin sexual experience in the back seat of a motor car.

Although my examples may seem extravagant, they are not. The fact is that, in a strange way, the side effects of technical innovation are more influential than the direct effects, and spread out in a civilization to transform its behavior, its outlook, and its moral ethic. For morality is an organization of life that grows spontaneously from activities, and not a formula taken ready-made from somebody else.

Of course we can foresee that certain modern technical developments will have profound social consequences. But we do not know what these are going to be. Let us take a simple example. I have no doubt that before my children finish the child-bearing age, say roughly during the next twenty or thirty

years, it will become a trivial matter for them to go to the doctor
and say, "We've had two girls; we want a boy." The doctor will
then go through a comparatively simple process (say, of centrifuging the husband's sperm, or whatever the method is by that
time—perhaps of centrifuging the husband!) and then he will be
able to guarantee, with 95 percent assurance, that the child will
be a boy.

We have no idea what the social consequences of this will
be. They might be manifold. It might suddenly become modish,
tomorrow, to have girls. The cover of *Vogue* or *Life* might
carry a picture of an alluring-looking woman, and all parents
will suddenly decide to have girls. After all, many parents
named their daughters "Shirley" not so many years ago. So having girl babies might be the kind of fashion which will sweep
over us.

On the other hand, what might happen is what they tried to
do in Italy and in Germany. They tried to encourage people to
have boys. They gave people private instructions about what
practices would (they thought) produce boys. Like all Fascist
scientists, they were singularly unsuccessful. But it is not out of
the question that if the Chinese knew the secret of producing
boys and were not producing enough, they would suddenly
switch to this practice. (And, you know, what we think about
Chinese militarism, the Chinese think about American militarism.)

If we discover that a simple method such as centrifuging
sperm increases the probability of having boys, the social consequences over twenty or thirty years may be massive. The obvious example today, of course, is the birth-control pill, which
already has had many social consequences, will have more profound ones, and will have some that are unforeseeable. For
example, it is already evident that the particular female hormones
which produce a good birth-control pill also keep the female
cycle going long past its present age. As a result, women of fifty

and sixty go on ovulating and, unexpectedly, have the look of younger women—fresher skin, hair, and eyes than we associate with their age. Consequently, the whole relation between the old and the young may change. Our society is geared to relations in which women think a man in his fifties still attractive, but men think a woman in her fifties unattractive. Now we may be within a generation of seeing that reversed.

Let me emphasize that while we do not know what the social consequences of a discovery will be, they do spring from very simple human motives. I have just been talking about men in their fifties being attractive to women. We also have the unusual situation now that men in their thirties and early forties are unexpectedly attractive to many teen-age girls. This is because the American government has chosen astronauts from that age group, and they have ousted the young Italian film star and the young Frenchman as objects of adoration. Who would have thought that the invention of space rocketry would have led to an age-shift in the image of the ideal man among many teen-age girls?

I want now to draw your attention to some foreseeable social consequences of modern technology. I have already said that, whenever people talk about genetic control in biology, they immediately ask such questions as, "Are we all going to be monsters? or supermen? and what's going to happen to kids like mine?" and so on. But, of course, that is *not* where genetics will be important in the near future. What will really happen is that genetics will begin to have its first influence in smaller ways. Our present control of the physical environment will expand to include control of the biological environment from small beginnings at first.

For example, the kinds of plants and domestic animals that we breed will be much more nearly tailor-made than they are today. I would like to give you two examples. First, let us ask, what is going to be the single greatest technological change in the physical sciences over the next twenty or thirty years? Rabi is the expert in this; I simply have the advantage that I am going to guess first; and my guess is that desalting of sea water is going to be the most important advance for overall world development. Because without this the whole complex problem of bringing underdeveloped countries to an acceptable level of economics, education, and political maturity is insoluble.

But if we propose to desalinate water so as to make it fit for drinking, we are setting a task which is really foolish, because at bottom we already have all the drinking water that we need. If we were to keep drinking water now only for drinking, and use the rest of the water for watering plants and other purposes, there would be no shortage of drinking water. So the obvious thing is to have desalinization processes which leave water as brackish as plants can stand it. And indeed a great deal of research in desert countries like Israel is directed to this.

As a match and counterpart to this, I guess that the single most important biological contribution to world peace will be to produce plants which grow effectively in quite salty water. This follows from what we all know about diminishing returns. If we are going to knock out all the salt in sea water, it is going to cost many times more than if we need only knock down 80 percent of the salt. So if somebody can come along and breed plants which can grow in 20 percent of the total salt content of sea water, we shall have the means to take a long economic stride. This is the kind of advance that biology in general and genetics in particular will make.

The other example I shall give concerns the breeding of animals. The potentially most useful animal that we lack at the moment is a sea animal which really harvests the sea efficiently. The countryside is full of animals which do a fairly good job of turning indigestible protein like grass into digestible protein like milk, eggs, meat, and so on. But in the sea, although there are such animals, they do it extremely inefficiently. If you take the smallest vegetable algae in the sea and think of the number of steps necessary before they are turned into a sizable fish you can eat—say, a sardine—the answer is discouraging. At present it takes three tons of algae to feed the small plankton which feed the larger plankton and so on until they make one sardine. This is a ridiculous ratio: three tons of algae to make one sardine. Nobody would breed cows or pigs if that was the ratio that you needed from vegetable to animal. So we are very badly in need of sea animals, particularly a scavenging pig of the sea, which have a higher efficiency than this. I have no doubt that we will breed them in the long run. There is no biological reason why we should not.

I propose the determination of sex, the pill, the breeding of plants to grow in very brackish water, even the evolution of a sea pig, as technical advances with major social consequences. We will get quite a different society, a different balance between city and countryside, a different ratio of power between Africa and the other nations, and so on, when these changes spread; and we might just as well face the fact that they will happen and spread. After all, George III in 1776 thought that there never could be civilization in a country where Indians were running about after dark. We must not be as short-sighted as he about the future—and a future marching on us much faster.

I have described first of all what modern biology is doing. It is the most flourishing new science in the world, and it is changing not only our attack on disease, but our whole conception of health and sickness. And second I have described how modern technical advances have many social consequences, both foreseeable and unforeseeable. Familiarity with these modern ideas is the best way of guessing these social consequences. I may know more about biology than many of you, but about social consequences we all start equal. If you know the facts, if you immerse yourself in the facts, you will be more far-sighted than the next man. You may come up with a practical or social gimmick before anybody else. But for this purpose it is necessary to be immersed in the new science, not to run out behind the field and start building computers when everybody else has already gone into the biology business.

Finally, I turn to my third subject and talk about the transformation in our concept of what life and the universe are about. Perhaps my best way of doing this is to say something irreverent. (I am sure that the one professional philosopher present will not object.) Philosophers often think that philosophy is made by philosophers, but the fact is that, for the last three hundred years, the most important changes in Anglo-Saxon philosophy have been made by people who have grown up in the sciences. And the reason is that scientists have had to invent new conceptions of what the world is like which in the end were absorbed into the common currency of philosophy. You have only to consider briefly the contributions of Thomas Hobbes and John Locke and David Hume, of Immanuel Kant and Ludwig Wittgenstein and Bertrand Russell, of Max Plank and Albert Einstein, to grasp my point.

When I was a young student I recall meeting a professor of philosophy who dismissed the whole of relativity with the criticism that it might be a very good scientific idea, but that curved space was a ridiculous thing to propose because in respectable linguistic usage the word "space" means flat space, and the word "curved" means the opposite of flat. Therefore, putting the two words together (he claimed), made the same kind of meaningless noise as if one said "a round square." No doubt my philosophical friend was pleased with this objection to general relativity. But the passage of time (and the urge of science) have shown that there is one thing wrong with his deadly linguistic thrust. There are no round squares, but physical space gives every sign that it *is* curved.

This illustrates how we have to accommodate our language

and way of thinking to the way that the world changes as science progresses. And we are about to see very profound changes of this kind. It is quite clear that we are going to start looking at the world in a new way. Problems of time and space, for instance, will take on a different air, for the simple reason that in biology today nobody thinks about the persistence of objects and the old philosophical problems of induction from present to future. Biology is not a science of objects (even of cells) but of processes. And thus we shall get from biology a differently directed outlook on the world, on the relation between objects and processes, and between past and future.

When we forget the professional philosopher and just think of human beings and their picture of the world, the most spectacular outcome of great scientific ideas is the way they have transformed man's image of himself. For in the end, people are more interested in people than in anything else. For example, look at the great human outcome of the Copernican revolution. When all the thunder had stopped, when Copernicus had died and his book had been published, when Johannes Kepler had fled from one country to another, when Galileo had been damned by the Inquisition and placed on the Index for two hundred years—after all that, what remained? People realized that the earth is not the center of the universe, and that man must not regard himself as specially created to be the lord of creation. Man's view of the universe and of himself in it was different; so that, for instance, Blaise Pascal could not have trembled at the silence of the infinite spaces if he had been born before Copernicus. People had not looked at the world that way before; they had not seen the heavens as something which dwarfed and overpowered them.

In much the same way, when the British Association had met in Oxford in 1860, when Bishop Wilberforce had stopped sneering and Thomas Huxley had stopped scorning his attack, when all the fierce pros and cons were over, what was the outcome of Darwin's *Origin of Species?* Human beings recognized that they are much more immersed in the totality of living nature than they had known. The greatest lesson that man learned from evolution is in seeing the animal creation and the world differently, and ourselves differently in them. It is not really relevant whether we see ourselves more modestly or less modestly—in different aspects, we do both these things—but that we see ourselves as a natural part of the animal creation.

In my own lifetime, the most important thing that has happened is that Albert Einstein adopted the strange word "relativity" for what he was talking about. The word, and the concept associated with it, swept the world. It is used by people who will never care whether the perihelion of Mercury behaves either as general relativity or as classical physics says it does. Instead, they use the word to express their new conviction that a great deal of behavior that had been thought to have absolute sanctions is really relative to a particular civilization. Physicists tear their hair; I remember how angry some of my colleagues used to become. They would declare, "But that isn't what relativity means in *science*. Why do they say that? That's simply comparative religion and comparative literature." My colleagues were right in strict scientific terms, but they were wrong about the human impact of the new outlook. The 3,000 million human inhabitants who interpreted relativity in their own way knew more about that.

I believe that as a result of recent biological work, we are on the threshold of an equally profound revaluation in human existence. And I ought to spend a last moment on this because it happens to be my passionate personal interest. I began today by talking about cells. We are all assemblages of cells. And even the brain, which I have rather neglected in this discussion, is after all an assemblage of cells. But if cells were all we were, then the obvious question is, "Why is there not a tidy assembly of forty amoeba sitting around in space and having this conference?" So we must be something else as well.

All right, what are we? We are mammals, we are primates. Agreed. But if we are merely mammals, even primates, if what is essential to our meeting here is part of their body functions, why is this conference not neatly staffed by forty cows or forty chimpanzees, quietly sitting around this table, having this discussion? There must be some residual that we have and that the mammals (and even the primates) do not have. It is now a fascinating question to ask, In what does this residual lie? What makes us animals and yet specifically human? What makes us able to sit here and make these notes and noises when even chimpanzees, who are really very intelligent, could not do it and would have no interest in so doing.

Similarly I am interested, in an anthropological sense, in how the human race came to be what it is over the last one or two million years. If we look at the skulls of those half-men of a million to two million years ago, we see that we have bigger brains. We have roughly three times as large a brain as those australopithicenes had. The human brain weighs about three pounds, their brain weighed about one pound. But the really interesting point is *where* our brains have developed.

We have, broadly speaking, developed the brain at three specific loci. The extra two pounds that we have put on, like all putting on of weight, is not spread all over the brain but is spread at very specific loci. The first locus is the speech center, the second is that area of the brain that is devoted to hand manipulation, and the third area is made up of the frontal lobes, whose exact function we do not yet exactly understand. But these great frontal lobes of man, which no other animal has, almost certainly have to do with what I would like to call imagination, that is, with our being able to construct hypothetical futures and to match these with things we want to do. So there are these three areas of brain development: the speech centers, the hand-manipulation centers, and the frontal lobes, which probably go with foresight and hindsight, with dreaming and imagination.

It is this kind of study which, I believe, has a universal bearing on how we shall think of ourselves fifty years from now—and how, indeed, young people are already beginning to think of themselves. I believe this will be enormously illuminating and fruitful. We are about to take another great step in awareness of the human self. We shall no longer stop at saying to ourselves, "Why, we're only a small island in space," as Copernicus implied; or "We're only another animal," as Darwin implied; or "We're only an accidental configuration of habits," as relativity appears to imply. For the first time we are coming to a point where people will be able to say to and of themselves, "Why, what makes us remarkable is this." And we shall be able to put our finger on some of the highly specific skills and activities that make us human. At last we shall have the right to say (the right given only by knowledge), "If we do so and so, if we behave thus, then we are fulfilling the human destiny. But if we ignore these gifts which evolution has given to man alone, then we really ought to have been born a cow (or at best a chimpanzee). We have no right to claim a human destiny if we refuse to be a human person—if we only use those parts of our make-up which a cow has got as well."

OPEN DISCUSSION

QUESTION. Would you enlarge upon your remarks about desalinization? Why do you think power relationships would be changed by a successful desalinization process?

BRONOWSKI. Let's take a very simple case like the Sahara. It is clear that somewhere around three thousand years ago, the Sahara was quite a livable part of the world. What the Sahara

lacks, at the moment, is water. There isn't anything else really
wrong with it. The old civilizations were invaded by sand and
did not resist those incursions. If the Sahara, in fact, had a water
supply and people began to build, as it were, islands of civilization in it, De Gaulle's dreams of French power would take on
quite a different shape. The importance of the oil and natural gas
deposits which have been already found there would be quite
different, if the Sahara were not an appendage of metropolitan
France but were, in fact, the main civilization, and metropolitan
France became the appendage. It's this kind of shift that I have
in mind.

QUESTION. What are the theological implications of what
you say? Are biologists hunting, now, for what others would
call the soul?

BRONOWSKI. Questions about God are easier to answer than
questions about man, simply because people don't think about
Him so much. The only time I ever heard a person asked the
question on a public occasion—"Do you believe in God?" and
who replied, "Of course not," was the biologist Francis Crick.
He is a rather devout unbeliever. How far this is characteristic
of biological scientists, I don't know. I would say, on the whole,
that biologists have been inspired by two things, and both of
them are connected with destroying an artificial sense of mystery. People like Francis Crick have been very much motivated
by people like the physicist Leo Szilard saying, "Look, I just
don't believe that biology is as mysterious as biologists have
made it out. I think that if you go for some specific thing, you'll
be able to find specific answers." And they turned out to be
right. People like this have wanted to destroy that kind of candy
philosophy with which a great deal of biology used to be so sur-
rounded. The attitude which says "Don't let us touch it. It will
all vanish, like a rainbow under our fingers."

With this comes the realization that, yes, human beings are
very different, but generalized remarks like "the soul" don't

really answer this. Theologians have, on occasion, disputed about whether an elderly lady who wants to be reassured that her pet dog is going to rejoin her in the hereafter is to be allowed to think that he has a soul. I am not being irreverent about this. This is what does agitate people. Now, on the whole biologists are not so much interested in whether the dog has a soul as in what makes the dog and the human different.

QUESTION. Are there, biologically speaking, any significant differences between the races?

BRONOWSKI. This is a fascinating problem. We don't really know how far the differences between any two people are genetic, and how far they are educational. The problem of what used to be called "nature versus nurture" runs through everything in biology.

Take a very simple problem. Rabi has won the Nobel Prize, but you haven't. Now, we do not know whether, in any circumstances, Rabi would have turned out to be a cleverer man than you, or whether, in his case, it was the whole educational thing—interest, and the family interest—or whether, in fact, it is a miraculous combination of two factors which just fit right and make a particular person very special. When you talk about a comparatively direct thing, that is, the difference between two people, nobody takes offense.

But if the two people had different color, then you're not allowed to say anything. So the whole question of whether the Negroes have the potential that white people have, how far their handicaps are educational and cultural and how far they are genetic, is more or less out-of-bounds. One thing is certain: at the moment, Negroes, Mexicans, Africans, and many others suffer such great cultural handicaps that it's pointless to talk about their native ability, because we aren't anywhere within miles of testing it.

I have no doubt that there are generalizations which can be made between races. If you can say that the skins of Negroes are darker than the skins of white persons, then there is already a genetic generalization between the races, and there must be others. But how far they bear on achievement is an insoluble question. It's an insoluble question for a reason that I'd like to draw your attention to. Read a book by Virginia Woolf called *A Room of One's Own,* in which she tries to answer the question as to why Shakespeare was a man and not a woman. She says, "Well, there may be something about men which make them capable of being Shakespeare. But we haven't really tried, because," as she says, and she's speaking about her experiences as a writer, "women have never had a room of their own to go into, to write their plays in." This is the basic problem. We haven't been able to test the genetic differences because we haven't equalized the part which environment and culture contribute.

DISCUSSION: I. I. RABI, JACOB BRONOWSKI, AND FRANK TANNENBAUM ON RETIREMENT AND SECOND CAREERS

QUESTION. Dr. Bronowski, won't all the biomedical advances you have predicted create a problem in employment, especially at executive levels? If people are going to remain healthy and active, we're going to have less turnover in personnel, so that we're going to have a little trouble in getting new ideas into the system as fast as we'd like to. I know that, in my company, I often wish that everybody had to retire at fifty-five and give the young fellows a chance. I'd hate to see it extended to seventy-five.

BRONOWSKI. Yes, it is a trend, at the present moment, to think that you ought to replace people who are fifty-five or sixty, in business. But this is going to change. Our view is going to change entirely.

RABI. It is not going to change the course, though, of early retirement. With today's rapid increase in knowledge and changes in environment, I think people are becoming obsolete after ten or fifteen years on the job. Maybe you could retire people earlier from one field and have them go in for a retread. A man could then be starting a new interest, a different career.

BRONOWSKI. I think Rabi's remark about retreading is absolutely right. Let me speak to you very frankly about people like myself and Rabi, who realize at the age of fifty or thereabouts, that we are still full of vigor. But the fact is that at that age we are never going to have a new idea in our own subjects. I can do mathematical problems, still, that many of the younger people either can't do or carefully avoid doing. But I'm not going to have a new mathematical idea. I grew up with mathematics in my twenties and my best work was done then. I'm through with it.

But it's very strange. At this late stage you can learn a new subject, get a new look at it, and have something to say about it. You don't have something to say in the way that a young man does, but you have other things to contribute. During the war I was thrown into things that I hadn't understood anything about, and I found them quite interesting. So I believe that one of the most important things about keeping ourselves mentally alive will be exactly this notion—that, in your forties, you go to a new job, whether in the same company or another company. You really ought to start at something which makes you learn afresh.

You heard me talk about biology today. What I've learned about it, I've learned in the last five or ten years. I speak to you like a man who has just come out of college because, so far as biology is concerned, that is my status.

QUESTION. Dr. Rabi says that every ten or fifteen years a man should change jobs. How many times can a man go through this cycle of lessons in his lifetime? Five times? Four times? We would find ourselves in a disturbing set of circumstances if a man had to change jobs every ten years.

RABI. That could be very interesting—one period in a man's life for education, one period doing maybe two jobs, and one period in his life living on his affluence and his thinking. It depends on how your society is organized. It seems to me that the process of education needn't be as formalized as it is. You go to school while you're working. You use evening schools. You combine your work with a constant education. I saw something of this in the Soviet Union, very much what we're doing today. Almost a perpetual school for people.

But in universities, I think the idea of tenure is pernicious. No contract should be for longer than ten years. The superannuation on the part of the faculty is fantastic. Furthermore, there should be a continued process of education. They should be taking in one another's washing, not only keeping up with things but broadening and deepening their interests.

I think that what most universities are doing now is sad, because in most institutions I've seen, the graduate students very often know a lot more than the faculty.

COMMENTS. I think the retirement problem is the vexatious one. If what Professor Bronowski tells us is true, we're going to have more and more old people who, for reasons of the job market—not only obsolescence but many other reasons—have retired.

RABI. I think we have a funny idea about retirement. Look at the advertisements for annuity funds. A man retires, and what happens? He goes off to Florida, leans back in a houseboat, and he fishes. How long can you last at this sort of thing? It's completely unrealistic. We need quite a different program for retired people, and we shouldn't use the word "retired" about such a man. He's just become a gentleman of leisure. It used to be a perfectly good thing to be. In the last century, an English gentleman of some means wasn't asked "What are you doing?" He did all sorts of things. He lived on his income. We have to revive that notion, and not feel that he must be in there whacking away at the American dream. Unless he goes into the office in the morning, fights his way through crowds and what not, and gets into a rat race, he isn't living. What nonsense! There are a number of things a man does, and can do, and will do naturally without this opprobrium of being on a shelf. He's not on a shelf. He's really there. The others are on a treadmill. There will have to be a different point of view. And once we get to that, the problem will be solved.

BRONOWSKI. I just want to support a point of Professor Rabi's. We have hitherto gone on the assumption that as the amount of work in proportion to the population shrank, we should have a shorter working day. Many people have the conception that in twenty years time the working day for the executive will be four hours, that older people won't have to work at all, and so on.

Now, I think that Rabi's absolutely right. This is not what's going to happen. What's going to happen is what we already know happens. The business executive today works exactly as many hours as his father did. And the man on retirement lives on the surplus. He doesn't halve his working time and live on the surplus, the other four hours of the eight. No, no, the surplus goes into other things—war, atomic development, taxation. But what we're going to do is to put this surplus into exactly what Rabi says—into what you might like to call pensions or retirement or so on—and that is, into allowing people to become gentlemen of leisure, doing their own things. We've reached a point where the surplus will accumulate, and can't be used for purposes which have nothing to do with shortening the working time of the younger people.

QUESTION. Suppose we use the term "renewal" instead of "retirement." We speak of a generation that is going into a second career. Is it a man's company's responsibility to send him back for renewal? Is it his? Does he have to provide this out of current income? This, I think, is a critical question. I think most men resist because they can't afford to retire and they can't afford to retread themselves. Perhaps this is a question of acceptance by society of a multiple career. Do you think they'll make allowances for this, or evolve something?

RABI. The "they" you refer to is you. Will you be willing to devote tax money to this?

TANNENBAUM. It seems to me a person's professional life and his personal life ought to go together. I think it's a great mistake to separate the two. There ought to be some way worked out so that a man's professional involvement, either in business or in a university, can continue as long as he lives. As he got older and ceased to be able to perform as he did when he was a younger man, he would always continue to be responsible in this relationship to the institution in which he had his career.

RABI. You believe in tenure.

TANNENBAUM. Call it tenure, if you want. The idea, it seems to me, is similar to that in the religious orders, like the Society of Jesus or the Dominican Order. One of the reasons why the Dominican Order is such a great institution, not only a religious but an educational institution, is that once you become a member of the Order you remain a member for the rest of your life. When you cease to be competent in one area, you carry on in another field. You always have a home; you always have a responsible range. You're never treated as if you're nobody; you're never completely abandoned.

RABI. But that's not the kind of society we've been building.

TANNENBAUM. I'm aware of that. What we're talking about is a mass problem of retirement. And what I'm saying is that if we're going to face that problem, we ought to have the ingenuity to devise a social structure that would make it possible for an individual to feel that he had both a professional and a physical home as long as he lived. Under those circumstances, you can build a great university, you can build an ongoing business with loyalty and attachment, and you can build a happy life.

I'll just make one more remark. I had a very close, loyal, and devoted friend at Columbia, who was a very great scholar. He's dead now. Ten years before he retired he said to me, "Frank, what do I do when I retire? Where do I put my books? Where will I work?" Ten years before he retired, he was already psychologically possessed of this sense of insecurity, a worry that his creative professional career would come to an end. You can't really be creative if this is the kind of thing you're obsessed about.

I'm saying to you that you can't run a great university unless you identify the person, physically and professionally, as well as personally, with the institution. I think that's probably true of an industry. The problem is, how do you devise a system so that you have no problem of retirement? You may have the problem of sickness, but no one needs to feel that on a given date he will suddenly find himself cut off from his friends. Don't forget that, when you retire, you no longer have a business telephone, you no longer have an office. Yesterday, to your students and colleagues, you were an important person; today, because you're retired, you're no one. You have no place to go.

BRONOWSKI. I quite agree with your scholar friend that this is a very important thing. But, you see, there are two solutions to this. A university might do either one of two things: it might provide these people with the amenities, among which a secretary and services are a very important part. Or it might take another line, which the army has taken, in which you are forced to retire early. But you are forced to retire so early that the comparatively modest pension on which you retire gives you a good start on a new career.

This kind of thing has happened in England. Now, you go up and down the countryside of England, and the pubs are managed by people who used to be boxers or football players or army or RAF officers ten years earlier. They've retired with a modest pension. Their first lives are over at thirty-five, if I may use the figure bandied about over there. I think it's a premature figure, but let's use it. In a way, being retired at thirty-five, with a modest pension, gives you a much better start on a new career than being retired at fifty, with a higher pension, and being no longer able to start.

Suppose I were running a business in aeronautics. I might say to myself, "Look. We know perfectly well that none of the presidents and vice-presidents are any good beyond the age of forty. So let's make an absolute rule that at forty, everybody has to go." And we would give them not, of course, the pension they would get at sixty or sixty-five, but quite a handsome pension. You know, the way the army retires its captains. If you don't get beyond the rank of captain, you have to retire at forty, and you make a new start in life. I've employed many captains and army people as a result of that policy. So it isn't true that an early retirement age is a great handicap, provided you have enough to tide you over this period of starting a new career. The worst retirement age is a sort of in-between age, when you have too little money to live on as a gentleman of leisure, and you're not young enough to start something fresh.

I think that, in the future, we shall see quite a bit of diversity in this. But I think we shall see quite a few firms which will say, like the army, that at quite an early age, a certain man is no longer useful to them, but that, on a modest pension, he can begin a new career in something else which is quite useful.

QUESTION. But you just said that you're going to make a man have a greater capability at a more advanced age than he has today. Well, then, why retire him at thirty-five rather than at seventy or seventy-five? Perhaps he can be useful that long in his job. Why don't we look at it that way?

BRONOWSKI. Because there are other considerations. There's

the whole question of the stimulus to the man of novelty, you see. And I would guess that, if you're working in the aircraft industry, you don't want men of very ripe experience. You are anxious to form good openings for young people, because it's at that age that they have very good new ideas.

RABI. I agree in principle. It would be a good thing for executives, for example, to retire early. When the children are grown and off to school, they should move back to the city and, for goodness sake, go into politics. We really suffer, in this country, from the fact that people with administrative ability, with means, education, and so on, stay away from running the government, which is done by a bunch of fourth-class hired hands. It really would be wonderful if such people would retire early, go into politics, take a serious public interest, and become really useful citizens, instead of hanging on. They'd get power, too, those who had a real drive for power.

Scientific Revolution and Cultural Continuity

A Comment by DAVID SIDORSKY

THE CHARACTERIZATION of our environment as an "environment of change" probes a latent conflict within major aspects of our cultural tradition. On the one hand, practitioners of humanistic activities—of literature, art, or philosophy—have often claimed that their results are unchanging or perennial, if not eternal verities or values, then at least enduring relative to the developments of new technology or to the discoveries of the sciences which have distinguished our environment as an environment of change. On the other hand, those who understand contemporary physics and contemporary biology, like Professor Rabi and Professor Bronowski, suggest that the impact of these sciences is not restricted to technology or to science but has been, and will be, of revolutionary significance for our society, including first, our human self-consciousness, second, our morality, and third, the conceptual framework or categorial perspective with which we organize our experience.

In the muted confrontation between the so-called two cultures which has taken place at the Conference, I was intrigued by the anecdotes which cast the philosopher as the traditionalist "straight-man" to progressive science. They reminded me of yet another scientific revolutionary, Freud, presumably a resident of

DAVID SIDORSKY IS ASSOCIATE PROFESSOR OF PHILOSOPHY, COLUMBIA UNIVERSITY.

neither culture, who wrote illuminatingly on the function of wit in argument, which in this case seems to be the release of some of the repressed tensions between philosophy and science. Those anecdotes told us about a Göttingen professor of philosophy who denied the possibility of atomic fission since linguistically an atom meant an indivisible entity, and of a Cambridge don who rejected the curvature of space since, conceptually, "space" was flat. Shades of that ancient and noble literary tradition in which Plato ends his internship as a philosopher-king by being sold into slavery or Aristotle completes his tutoring of the young Alexander in the virtues of finitude, power sharing, and the small city-state with Alexander's embarkation on an infinite imperialist venture! Now, though my comment is motivated by a desire to defend the professional honor of philosophy, I concede the historicity and, more relevantly, the force of the anecdotes. The scientific thought of an Einstein or Bohr has shred some of the philosophical knots regarding the nature of space or of the atom as effectively as Alexander's sword cut the Gordian knot.

And yet, without endorsing the antiscientific conservatism of the German atomist or British space-man, I should like to advance several reasons why changes in our morals, in our self-consciousness, and in our conceptual framework are, to a marked degree, immune from the rapid rates of change of biological or physical science or technology. After all, it is precisely because it is true that the meaning of the word "atom" does not determine the scientific possibility of atomic fission, that the conceptual reformulations of science, say, that time is a dimension, do not change the rules governing tense in English grammar.

SCIENTIFIC PHILOSOPHY

Let me however first hasten to wave the flag of truce between philosophy and science. However grave and disturbing for our society is the ignorance of humanist and philosopher

about the methods and results of contemporary science, the
major activity of much of modern philosophy has been an heroic
effort to restructure philosophy as scientific philosophy.

This effort is coeval with the rise of modern science in the
seventeenth century. Witness Spinoza developing an ethic which
is presented as an axiomatic geometry of bodies in a deterministic
universe. Indeed, many modern philosophers have not just
sought to wave a flag of truce, but to flaunt the colors of revolutionary science. Witness Descartes, who warns himself in a characteristic bit of *monologue intérieure* that prudence dictates his
witholding from publication the manuscript of his *Discourse on
Method* in the year in which Galileo has been censured. Descartes then concludes that he must cast prudence aside. This risk
is not taken for the glory of philosophical truth, which is a vain
delusion—philosophers after all have espoused incompatible
views since time immemorial with the conviction that each has
philosophical truth—but because the fruit of Descartes' method
is a rational and progressive method of discovery for physics and
biology. Biological science, not philosophy, is worth the risk of
martyrdom, since, as Descartes prophesies (and if I followed
Professor Bronowski, he confirms), its stakes are the elimination
of "l'affaiblissement de la vieillesse" and perhaps even of "la
facheuse habitude de mourir."

Twentieth-century philosophers have also sought vigor-
ously to bind themselves to the methodology of science. The
charge most often asserted against contemporary philosophy by
its critics is the charge that philosophy has abdicated to science
its task of comprehensive or synoptic speculation of the "world
as a whole" and has thereby forfeited its role as a central vision-
ary guide into the nature of first and last questions. I believe the
charge of abdication to be a just one although the consequences
may be no more lamentable (to use Bertrand Russell's piquant
illustration) than the loss of vision, guidance, and imagination

which took place when astrology was transformed into astronomy.

My own preferred account of the implications for philosophy of a wholehearted acceptance of the methods of science as the only way to understand the nature of things is the following, alas, only partly correct analogy. Just as the portraitist who had striven for representative verisimilitude has discovered after the invention of the camera rendered him technologically unemployable that his true métier had always been, though unknown to himself, the arrangement of elements of color, texture, line, and expression in the guise of a portrait, so the philosopher, liberated from the necessity of scientific pretentiousness, can now examine the conceptual issues in the sciences, in common sense, or in any stratum of language—which always were the distinctive center of his philosophical concern even when he was involved with empirical inquiry or literary invention. Whether in this manner or another, the demand that the full implications of scientific accomplishments be recognized by philosophical enterprise has become an internal demand of philosophy. To that degree, the strictures suggested by Professor Rabi and Professor Bronowski about philosophy have received, what Norman Thomas, among others, lamented as the most effective treatment for criticism: adoption of the critic's point of view as your own.

This peace offering having been given, there remain, however, a number of questions in which the general point of view advanced by Professors Rabi and Bronowski, as I interpret it, would conflict interestingly with one philosophical view of the reality and importance of cultural continuity.

HUMAN SELF-CONSCIOUSNESS

It is a truism that modern science has radically changed the ways in which man views himself. Yet the proof of that truism, which should be trivial, presents serious difficulty. The sharp sketches of the "before science" and "after science" images of man which are needed to exhibit the contrast blur upon reflection.

The "before" scene is familiar enough. Man is cast as aware of himself as a sovereign creature possessing the light of Reason. Reason places him in the great Scale of Nature at the critical midpoint between animal and angel. His home is a planet which is both the dramatic and geographical center of a single, finite universe which stretches from Hell below upward through a series of luminous spheres to the Divine Heaven. The "after" scene is a palimpsest of images but they presumably mirror the Darwinian vision of man as risen ape, not fallen angel, who finds himself not "at home" in a post-Copernican universe of overpowering dimensions without any hierarchical structure.

A whole anthology of poetry could be presented by way of documentation. Having recently read Douglas Bush's *Science and English Poetry*, I cannot resist quoting for you some of the lines. For the new physics, from John Donne, whom I am informed by Professor Bush accepted Tycho Brahe's, not the Copernican astronomy:

> And new philosophy calls all in doubt:
> The element of fire is quite put out;
> The sun is lost, and th' earth, and no man's wit
> Can well direct him where to look for it.
> . . . .
> 'Tis all in pieces, all coherence gone . . .

For the post-Darwinian biology, it was, aptly enough Aldous Huxley, a descendant of Darwin's bulldog, Thomas Huxley, who wrote in "First Philosopher's Song":

> But oh, the sound of simian mirth!
> Mind, issued from the monkey's womb,
> Is still umbilical to earth,
> Earth its home and earth its tomb.

The difficulty with the thesis of the significance of biology and physics for self-consciousness, however, is not the absence of evidence for the thesis, since there is abundant evidence, but the equal wealth of negative evidence. For every line like Donne's which is occasioned by the new astronomy can be matched by one which shows man at ease in the Newtonian world of "Great Nature's well set clock." Further, there are many statements of cosmic anxiety from the pre-Copernican past, derived from the sense of the immensity of the "old foundations of the earth" which the poet defied Job to "lay the measure thereof." Every statement of man's sense of continuity with animals stemming from the new biology can be matched with one which praises the uniqueness of human estate in the evolutionary struggle. Further, there has been an age-old awareness of man's animal essence.

Since an empirical hypothesis is not at issue, the future results of science will not provide the solution. The Aristotelian claim of human rationality as the unique and precious potential for human fulfillment, which Professor Rabi reaffirmed here as the belief of Einstein, will not be expunged by mathematical discoveries of the formulation of rationality as Turing machine computability. The despair over "irrational man" or over his "returnability" to dust will not be eradicated by that genetic research which would show, as Professor Bronowski has illumined, how complexly organized and richly bonded that particular dust is.

No one can deny the role of scientific data on the nature of man and of the world in forming our image of man in the world. The information, however, in a sense, overdetermines the result. Between that data and our self-image come the mediating set of attitudes, common-sense knowledge, and perspectives which are crucial. I do not know how the perception, the selection, or the structuring of the data takes place. I assume that crucial features in our self-consciousness are not contained in the scientific information. They are partially in history. It is Auschwitz, not Darwin, which heightens our image of ourselves as "irrational man." Their location is, as Frost has written (in an apparent direct reply to Pascal's pensée "the eternal silence of these infinite spaces frightens me") "much nearer home":

> They cannot scare me with their empty places
> Between stars—on stars where no human race is.
> I have it in me so much nearer home
> To scare myself with my own desert places.

SCIENCE AND MORALITY

There is an anecdote which the "moral traditionalist" might use to counter the view—more correctly to moderate the emphasis—so well developed at the conference of the revolutionary implications for morality of the new technology. H. G. Wells is said to have found fault with the burgeoning crop of science fiction in the early years of this century. Although its scientific and technological extrapolations were superb, Wells criticized its lack of psychological inventiveness. Interplanetary travel and communication, explosive rays, and wonder drugs abound, but the human aspects of the plot are bounded by quarrels over a girl or over gold, and marked by petty resentments or vainglorious ambitions. Wells suggested that the understandable reason for the discrepancy was the extraordinary difficulty of cultivating our moral imagination. For it was obvious to Wells that a society which would have achieved such high levels of science and technology would also have transformed beyond recognition human motivation and character.

In retrospect, a significant portion of the scientific achievement projected at the turn of the century has been achieved. The plots, however, not only of our updated science fiction but crucially of our political and international life, remain starkly old-fashioned. In the profound lyrics of the Bogart-Bergman movie theme song, "It's still the same old story . . . As time goes by."

Regrettably, neither anecdote nor lyric probe the harder issues. Professor Bronowski has, after all, suggested some ways, not as many as I would have suspected, in which the coming biotechnology will affect our manners and our morality. Since I agree that the substance of morality is more often to be found in the everyday patterns of manners than in the alleged melodrama of ultimate moral choices, I agree that these changes in human relations to which Bronowski referred—that women will age more slowly than men and therefore older women will more often marry younger men, for example—are of cumulative moral significance.

Further, even if the moral plot, so to speak, of our political or economic life is unchanged, a striking shift in the technological resources available to the players may be of moral relevance. Some suggest that the morality of warfare has been affected by the new scope of military technology; perhaps the morality of distributive equality will be affected by a revolution in productive resources. Thus the argument is that even if the moral scenarios of the century are not significantly different from those of an earlier age, the quantitative leap afforded by technology has shifted some of the moral dimensions.

Although I am skeptical about this argument, I retreat to the rear line of my conservative defense of the continuity of morality in an age of scientific and technological achievement— the continuity of moral concepts. Unlike terms such as "atom" and "space," which are redefined relative to the regnant scientific theories, terms like "free" and "just," "right" and "good," are not. It is true that the concepts of morality vary in meaning in different languages or in different societies at different times. It is even true that redefinitions of these terms in new systems of political and social thought may eventually and partially transform their meaning. Still the control of these terms is given by everyday life and speech. The changes of meaning which have taken place through their history pale in significance before the sheer continuity of their use. How else explain how children for thousands of years can capture fully the moral force of Nathan's quarrel with King David or that generations of students grasp the nuances in the dialectic of Socrates which exhibits that "justice" is not quite the same as "telling the truth" or "punishing the wicked" or "asserting one's interests"?

CONCEPTUAL CONTINUITY

The permanence of our conceptual framework is not uniquely reserved to moral concepts. Though the sciences grow through a continuous redefinition of terms, the conceptual apparatus with which we understand the major features of the physical or animal world remains stable. This can be traced in the familiar and pervasive functions of a variety of languages. We identify common objects through substantive nouns; articulate these objects as definite or indefinite; modify them with adjectival properties of color, shape, or size; point to them with demonstrative pronouns, and so on. (Perhaps, as Santayana speculated, the elementary concretions in discourse and perception which constitute common sense and are exhibited in ordinary language are basic residues of human neurological evolution.) Let me examine the relevance of this thesis of conceptual continuity for our main theme, the understanding of social change, by developing a somewhat fanciful but not necessarily false illustration.

At one of the first meetings of the Seminar on Technology and Social Change, William Baker, of Bell Laboratories, reported some remarkable engineering achievements involving the measurement of fantastically short spans of time, on the order of trillionths of a second. Even if it is true that I am making use in my transistor radio or telephone of some of the fruits of that technology, the consequences of that discovery does not affect my way of measuring time, or of organizing my time, or of being aware of duration.

Compare that with the observation made by Lewis Mumford that the widespread introduction of reasonably accurate clocks in the early Renaissance was crucial for the development of technological, capitalist economy. It did not significantly affect the "product-mix" of the economy, but it helped change the way the working-day was organized, and made possible much greater management and control of input and output. The ubiquitousness of clocks inside walls may plausibly—I am not sure how true this is historically—effect radical social change in an agricultural society organized around those fairly reliable clocks: daylight and seasons.

Now, a third way of marking the passage of time is by the use of tenses in language. It has often been pointed out that a different attitude to experience is conveyed by a language in which events are discriminated as "complete" or "incomplete," rather than as "past," "present," or "future." Edmund Wilson has complained persuasively of the difficulty of arranging appointments or asserting standards of accomplishment in a society whose language merges the present and future into the sweet-by-and-by of incompleteness. The obvious candidate for contrast in this *jeu d'esprit* is Latin as the language of magnificent military and administrative organization of the Roman world, with its sharply demarcated pluperfect, perfect, imperfect, present, participial present, future, future perfect tenses, fitting into the rungs of measurable activity. Interestingly enough, one Hungarian scholar has sought to prove that so sharp are the verbal demarcations that the principles of calculus could not have been formulated in ancient Latin.

The limit to linguistic conceptualization of time processes would be to try to imagine a language in which there was no way at all of ordering sequences of time. Perhaps isolated actions could be names but there would be no words to construe the "before" and "after" in a process. Thus it would be as sensible (conceptually only, I fear) to conquer before one sees or to see before one comes, as it is first, to come, then to see, and then to conquer. It was Kant who pointed out that in such a language not only would mathematics be impossible but the discrimination of experience required for a rational ego would not hold. Generally, it would seem not too far-fetched to say that to be rational requires some sense of time sequence.

The function of this lengthy illustration is to lend force to the idea that there are different levels of conceptual activity and appropriately different rates of change. We may often encourage extremely rapid technological innovation. Major corporations proudly report that they are manufacturing only products which did not exist a few years earlier. This is however compatible with arguing that changes in the organization of the individual's life cycle—early retirement, career retraining—should be introduced very slowly. We should perhaps be willing to accept significant economic costs to slow down rates of change in these potentially self-defeating uses of human resources. It is also compatible with arguing the view that some aspects of the human condition, including our sense of self, our moral perspectives, and our categorial framework, are relatively unchanging. Our educational or welfare system, for example, may require a heightened awareness of the fairly constant pattern of human needs.

One thesis of our analysis of the implications of the new physics and biology is that our survival and our sanity depended upon our learning to be at home with, to adjust to, to control, to cope with the discontinuities introduced by technology. A slightly less urgent moral—less urgent not because it is less important but only because its achievement is probably easier and partly inevitable—is that our survival and sanity also depend on not forgetting that we continually cope with certain ultimate continuities of human existence.

If there are different appropriate rates of change for various levels of activity then the trick is, and has always been, to reconcile the elements of continuity and discontinuity. This is hardly a novel thought. John Dewey preached for several decades that the gap between our acceptance of rapid rates of change in the physical sciences and our resistance to the use of the social sciences represented *the* crisis of American society. My own sense of what is to be reconciled in our society is a different one. It is, however, only when we face up to what is to be reconciled with what, and how, that we can appreciate the depth of Professor Rabi's comment that in this area most men assert "deep truths" which were defined by Niels Bohr as something the opposite of which is also a deep truth. "Nothing ventured, nothing gained," I may say as I press the discussion, but I also realize that in this domain, "fools rush in where angels fear to tread." The only time there seems to be some point in asserting platitudes is when they are being contradicted or overlooked. My calling attention to the continuities of the human condition may be appropriate when developments in biology and physics seem to suggest that self-consciousness, morality, or conceptual perspective will be dramatically revised, as in a certain sense only they will, by those developments.

Since, however, I believe that the question of the historical continuity or discontinuity of our time cannot be empirically answered, why put the question or formulate an answer? The reason seems to be that in asserting continuity or discontinuity we are suggesting a framework for action, expressing a degree of confidence or of concern about the future. This question was recently put to Stravinsky, who replied pessimistically that the future belonged to discontinuity because of the revolutions in biology but then, I think, and I shall quote his answer in full, affirmed his confidence (at eighty-five) in limited, personal continuity. I thought this point emerged in discussion at the conference when Professor Rabi voiced his fears over nuclear warfare: we are not cheered by the continuity of balance-of-power politics as much as we are haunted by the threatening discontinuity of nuclear warfare.

There is, however, a logical, true, and appropriately nugatory answer to the issue of historical continuity. Logically, either we shall survive or we shall not. If we will not survive then there will be discontinuity. The only credible eschatological literature is apocalyptic. We do have, as H. G. Wells knew, the moral imagination for disaster.

If we survive, no matter how rapid the rates of change of biology and physics, then continuity will be proven. This necessary truth was one of the few agreed to by both Plato and Aristotle at the birth of philosophy on the ground that if all things were totally changing then nothing could be grasped. It is only as the result of our scientific achievement, no matter how revolutionary, is woven into the texture of our historical and human experience that it effects some measure of change. This provides the obvious reason why the utopian works engendered by the miraculous achievements of science never totally remake, in fiction or practice, the human condition.

Stravinsky quotes Yeats's dictum "Myself I must remake." I conclude with his reply to the question:

Interviewer: "Have eighty-five years strengthened your belief in the continuity of life and art, Mr. Stravinsky?"
Stravinsky: "It seems to me, on the contrary, that discontinuity must be only a short way ahead. Certainly in another four-score-and-five the clones bred by genetic engineers to the specifications of Bureaus of Human Uses will differ from me far

more than I do from, say, the astronomers of Fowlis Wester and Stonehenge (who, it now appears, must have known a lunar movement discovered by Tycho Brahe). I lack confidence in genetic (and other) utopias, as you see, and am tired of analogies based on successful demonstrations in frogs. But, then, the present has already begun to make me giddy—that discovery at Cambridge, for instance, showing that in certain insects the sense of time itself is encoded in two or three cells. (So, then, time *is* ontological?) But do not misunderstand me. Though I would refuse any Faustian bargain, I am far from content with myself. I simply want to go from here on out trying to do better what I have always done, and in spite of statistical tallies telling me that it must be getting worse. And I want to do it in this same battered but long-lived-in Identikit. Myself I must remake, Yeats's poem says. And so must we all."

Technology, Work, and Values

ELI GINZBERG

THE DISTINGUISHED American economist Thorstein Veblen wrote a long time ago on "The Danger of Taking the Lead." This referred to Great Britain's lead in the Industrial Revolution. I am glad that Sir Isaiah Berlin and Professor Rabi took the lead yesterday since they gave me a way into my subjects.

I asked Professor Rabi whether he follows the same predictability system in forecasting the weather that we use in economics, which is essentially that tomorrow's level of activity will be more or less like today's. He said that was about right. I agree with Professor Berlin that predictability in history, as in economics, is very complicated.

In his lecture, Professor Rabi talked about indeterminacy; it reminded me that at first economics was based on the Newtonian system. Consequently, neoclassical economics was a determinate system. When one of the forces is moved, everything else returns to equilibrium. More recently a series of ideas have been promulgated in which the outcome is less determinate. Keynes, for example, worked out a system of thought which indicated that the economy is not self-adjusting and that equilibrium at full employment is the exception and not the rule.

I will consider the manpower part of economics rather than the money-making part or profits. Manpower has one charac-

ELI GINZBERG IS HEPBURN PROFESSOR OF ECONOMICS AND DIRECTOR OF THE CONSERVATION OF HUMAN RESOURCES PROJECT AT COLUMBIA UNIVERSITY.

teristic which makes prediction a little easier. If we assume that people will continue to work until the age of sixty-five, then there are people alive today who will be members of the labor force in the third decade of the twenty-first century, unless there is atomic holocaust. As far as manpower is concerned then, we have more time to see things in perspective, and that may be of some advantage when we discuss change, the economy, and manpower. I think it was Professor Eiseley who pointed out that we do not live long enough to become as wise as we should. However, when one deals with such aspects of the economy as manpower and has become as old as I am, one has the advantage of some perspective. I may not have real judgment, but I have lived through a few cycles, a few episodes.

My experience goes back to the perpetual prosperity of the 1920s. I wrote a book called *The Illusion of Economic Prosperity* after prosperity turned out not to be so stable. We did a study on long-term unemployment in the 1930s and saw it reversed because of war mobilization. I remember discussions with fifty of the top economists in Washington in 1944 about the great depression that was going to follow when we converted from war to peace, but it never came. I remember the 1950s, when we were writing books on the shortages of manpower, only to run into an unemployment rate of 7 percent in 1961. I have been advocating an active manpower program during the last few years, and now we again have a shortage problem as well as some residual unemployment. Thus although we deal with a long-lived resource, it is a complicated field.

I want to make a few remarks about technology. Economists used to make the problem very simple; they used to stipulate that technology was a stable factor. They ignored technological phenomena, and simply made their analyses in terms of market mechanisms. Because they realized that if technology did change, and that then their whole system would become very complicated, they assumed that technology would remain stable.

This will give you an idea of the gap between academic economics and the real world.

There were, however, a few economists who understood that they could not study economics under such a stringent assumption, and they put the factor of technology into the middle of their considerations. Thorstein Veblen was one of these people. He argued that understanding of the economy is to be obtained by considering changes in technology; he even took this a step backward—to developments in pure science and in advances of knowledge in general. He understood that the basic capital of a society is the accumulation of knowledge; from it technology is derived and from technology the economy curves. He had some interesting arguments about the nature of private property. He said that scientific property is public property, and that is really the basis of wealth; he said that technology is an adaptation of scientific knowledge for the common good. These are simply examples.

There is also the theological doctrine of eschatology. People who prophesy doom are called eschatologists; they believe that the world is running down, or that we are going to blow up, or some other Jeremiad. I had a close association with an early version of technological eschatologists—they were called "technocrats" in the 1930s. They had a picture of the world in which everything was going to be different by virtue of technology, and the leader of that group was in the engineering building at Columbia University. Today, the cybernetics group, in their more extreme variant, remind me of the eschatology of those technocrats. They too feel certain that the world will be unsettled by technology and that tremendously rapid and fundamental changes are going to take place. I will come back to this point later on, but I want to identify this school of thought now.

Others believe the opposite. These belong to the optimistic tradition. They are the people who say, "Technology is the mainstay of economic growth and of prosperity and higher per capita income."

I shall try to make clear my understanding of where the truth lies between these extremes. The argument that technology can either be ignored or is relatively unimportant seems foolish. The argument that it makes for such orders of discontinuity that we cannot use the past at all also seems foolish.

Let me state some hard facts. If we use 1929 as one base line and 1963 as the other, certain very strong conclusions can be extracted from the experiences of the American economy. One is that technology, which has had a major impact on American agriculture, has been associated with very rapid declines in employment in agriculture. Soon nobody will grow food on farms because all agriculture may eventually become a laboratory technology. In any case, in the United States we now have a farm labor force of 4.25 to 4.50 million, which means that one farmer feeds about 16 other Americans in the labor force. Let us say, for our purposes, there are 75 million people in the civilian labor force and 5 million people on the farms. One farmer supports 15 other workers, and in addition he creates a surplus for overseas. Any reasonable interpretation of the farm situation indicates that there are still too many farmers. We could probably farm our farms with 2 million less. So the first impact of technology has been that it has reduced employment in agriculture tremendously and continues to squeeze it.

Let us look at mining. By and large the Appalachian problem is very much a result of technological advance. The American mines are advanced technologically, because John L. Lewis made a deal with management—he agreed to management's technological innovations if they would pay high wages to the people who continued to work. Many lost their jobs. We now have half the number of miners we had in 1929, and a much larger output.

Then, if we look at the hard core of the American economy

—manufacturing—which is the most interesting part, we see that the peak of manufacturing employment in the United States was in 1943; we returned to that peak only a couple of months ago. And if we break manufacturing employment down into blue-collar workers (those who work on the factory floor) and white-collar workers (the people who plan the work, control it, and maintain the machinery), the number of blue-collar workers is still below the 1943 level. At the same time, of course, output has been increasing quite rapidly. We can say, then, that it would be reasonable to assume that in agriculture, mining, and manufacturing, technological advance has been proceeding at such a rate that it is hard to conceive of any large additional employment in those sectors.

In construction, which is a so-called backward industry, productivity changes have not been quite so rapid. Although there has been an increase in employment since 1929, there have been a lot of technological changes in this field, for instance, in basic foundation work.

How then has the economy absorbed all the additional people? It has absorbed them in what we call the service sector. Part of the service sector is made up of the communications industry, in which there is heavy capital (and there, too, all kinds of economies have been made in manpower): the telephone industry, the railroad industry, the airplane industry, and others. Also in the service sector are retail trade, health, and education—where the use of capital is less prominent. Nevertheless, we have seen only the beginning of the impact of the computer on the white-collar employee. We are only beginning to understand what is likely to happen; we have not seen what the new technology can do, even with respect to blue-collar employment.

On Forty-third Street and Eighth Avenue, in New York City, there is a new completely automated parking garage where you can claim your car through a computer. The key is put in the computer and the car is brought out by a machine in six seconds. All the bookkeeping is done simultaneously. One man sits there—he can't even speak English.

With regard to employment, there is already enough evidence in the changes that American economy has been undergoing since 1929 to show that it may not be as easy to keep the balance among technology and employment and output in the future as it has been. Technology has moved from one sector of the economy to the other and it is now entering the service sector in a powerful manner.

Our heavy investment in research and development, which is now at about $20 billion annually, is a new phenomenon largely connected with war. This can be interpreted in two ways. Its immediate effect on the economy is not very great. Mr. Webb, who runs NASA, has had great confidence in what the $5 billion annually for space programs is doing for industry. If we look closely, however, we see that the money does very little directly for industry. On the other hand, we find that civil aviation, the electronics industry, and many of the developments in the medical area have been heavily affected by military research. Here is an indirect, slower, but potent effect on the economy.

Last year Congress passed what the President called a piece of "sleeper legislation," that is, a technical assistance program which will bring technological and scientific advances more readily to small-scale industry. It is much the same type of program as the agricultural extension service.

One important point about research and development programs is that although the demand for scientific and technically trained manpower in America has been greatly expanded, this manpower lives next to the private economy and is totally dependent on government funds. The extent to which our economy is a defense economy should not be minimized. In 1958, President Eisenhower did not want to ask Congress to lift the debt ceiling, and directed the Air Force to elongate its contracts. Within sixty days the presidents of the large universities involved and representatives of the aerospace companies which had government contracts were in Washington to tell the government that without the contracts they would be out of business. They did not have sixty days' working capital to keep their research teams going. That indicates the dependence of the sector of technical manpower on governmental expenditures. I am a very small research director at Columbia, and I do not have ninety days' working capital. My contract has to be renewed annually, or a lot of people will be out of work. The extent to which we are transforming the economy is indicated by the fact that two-thirds of all the scientists and development engineers in American industry are supported by nonprofit dollars. This is a new development. And this, again, fits into Professor Berlin's presentation of untoward consequences that were never planned.

The next aspect is that if we push science and technology we also push the skill-obsolescence rate. That is, we destroy a great deal of human capital by speeding new scientific and technological developments. A large number of jobs for unskilled and semiskilled workers are constantly being liquidated by our new machines.

Scientific and technological development affects both ends of the skill continuum. It makes people with the top skills more quickly obsolescent, and it often eliminates the jobs at the lower end of the scale.

Hannah Arendt said that fewer and fewer people are playing a significant role or any role at all in the economy. I say that the bulk of mankind never played a decision-making role in the economy. They were peasants and worked the fields, or they were simple laborers in factories. I do not think they wanted to play a more important role. I do not think they were trained to play a more important role. Today those entering the work force have had on the average one year beyond high school and they are more likely to play a significant role in the economy than in earlier times when the level of literacy was very low.

Moreover, I do not think all the simple jobs are going to be liquidated, because there are a large number of jobs in the service sector that must be done, for instance, waiting on tables; many jobs are supplied by the field of travel and tourism; a large number of jobs in hospitals are jobs for the semiskilled. The notion that we will liquidate all jobs for the unskilled is unfounded. The fact is that only certain kinds of jobs are likely to be liquidated.

A result of increasing skill obsolescence is that those whose professions require a long initial period of education, physicians, for instance, are the first who must continue their education. Physicians were the first group to move to a continuation of their education, because even though their original education was very elongated, it was not adequate. A parent with a sick youngster does not want him treated by the honor man of the Harvard Medical School who graduated in 1950 but who has not looked at a new medical book since then.

The university offers a particular advantage; it remits to its professors about 50 percent of their effective time to permit them to stay up to date. Most of us do not take advantage of the offer, but that is the essential contract with the university. Now, interestingly, American industry has no perception of the rate at which it makes its trained manpower obsolescent by failing to give them a chance to keep up with current research and development. It is important for a private employer to encourage this, because in a rapidly changing world, it is costly to use people for only five to ten years and then have them atrophy.

In a large corporation like Du Pont, people can be moved from the laboratory to sales or manufacturing, but it is not always desirable. The basic problem is to keep profitably active those people in whom society has made a heavy investment by their initial training.

There is another way in which technology influences the economy. If we think in regional terms instead of national terms, we see that new resources are opened up by virtue of technology; for instance, scrub pine in the South becomes a new lumber industry; or, as a result of the development of nuclear power, the management of the coal industry becomes alarmed and tries to lower costs. We see a tremendous circular impact.

The final point about technology is that most scientists and technologists are working with the help of nonprofit dollars. We have become a pluralistic economy; we used to think of the economy in terms of business and government. The government used to extract valuable profits from the private sector and, depending on one's political opinion, waste it or underwrite an essential function with it. Mr. Hoover's view was that the government would waste the money. The more up-to-date view is that government did some useful things that were necessary even for the expansion of business. But that is a very unsatisfactory model. Actually, we should stipulate a three-sector analysis—in other words, a pluralistic economy—government, private enterprise, and nonprofit institutions. The key nonprofit institutions are the universities and the health institutions, although there are many others. Government, of course, is also divided into three parts—federal, state, and local. And again there are circular relationships. For instance, industry gets its manpower largely through government and nonprofit institutions. It does not provide the basic education of its people. The great chemical, petroleum, electrical, and manufacturing industries depend on the universities to do their training. Industry picks people after they have been trained. The basic technology used in industry is primarily university-engendered, and many developments have been engendered in governmental laboratories.

Thus the notion that the dynamism of a society lies in one particular sector, for instance, in American industry, is simply not borne out by the facts. Let us consider our most dynamic industry—automobiles—which is the easiest example. We think of it as the quintessence of private enterprise. But the industry would not exist at all without a highway building program, and that is 100 percent governmental. One of these days, when the story of the automotive industry is properly written, it will show that the real genius of the people who ran it—after Mr. Ford worked out the problem of how to produce the cars on a mass basis—was their developing support at state and federal levels for colossal highway programs. It was the *sine qua non* for the development of an automobile industry.

As you see, these relationships are circular. But now, interestingly enough, an appreciation of their circularity is beginning to penetrate. Recently in Kansas, which is not a very radical state, an economic survey recommended that the state of Kansas establish some professorships at $25,000 each. Nobody in the whole government of Kansas earns that much. The establishment of these professorships was recommended to make it possible to attract technological concerns into the state once the University of Kansas had been built up. And that would be the basis of Kansas' economic future. When Kansas sees interrelationship between universities, technological industry, and general economic growth and development, that is an appreciation of cirularity.

Let us now leave technology and discuss work. The first point is that in the advancement and the development and the elaboration of science and technology, there has been a constant shift in the skill levels from brawn to brains. People used to get off the boats from Europe without knowing a word of English; they had never been in a factory, and yet they had work at once. A lot of recruitment was done on the docks of New York City, literally, people were told to report to work the next day. A typical screening device used early in the century was to ask a fellow to roll up his sleeves, and then look at his muscles.

It is quite clear that brawn is not the major demand in the year 1967. There are only a few jobs today which require muscle. René Dubois made a beautiful point about our changing patterns of adaptation. He pointed to the importance of good eyesight for fighter pilots in World War II. In those days a pilot had to have eyes in the back of his head. Now big planes with instrument panels put a premium on pilots with a certain amount of astigmatism.

In any case there is increasing pressure to plan and coordinate large structures, and this is one reason that management works so hard. Many decisions must be made at the top—decisions which are of compelling importance to the organization. The key decisions concern investment, new products, the shape of the research and development programs, and the location of plants. The key decisions guide the company, and they are not easy to decentralize.

Machines can take over only routine work. Therefore the computer may eliminate increasing numbers of those in middle management—those who today are concerned with inventory control and sales forecasting. Many of these problems can be solved by machines, and there will be fewer people involved in processing materials on the first level. There will always be people at the second and third levels.

I have a personal bias: I believe that in many organizations there is one person in six who actually carries the work of the organization; the other five go along for the ride. The five are there; they do a day's work; but they are unwilling to risk taking initiative. This is another reason that certain members of the management group work hard.

The next question is, what is likely to happen in the service sector of the economy with respect to work? Here we come to important qualitative dimensions. I find it strange that one of the major chemical companies still has the same time requirements for the research laboratory that are applied to the manufacture of bulk chemicals: the research scientists are supposed to punch a clock at nine and stay until five, and the laboratory is closed on Saturday and Sunday. The company treats its scientific and technical personnel just as it treats a man who is hired to work on production. Of course, a company can insist that the research staff come in from nine to five, but unless there is some way of estimating their productivity, reliance on the time clock is ineffective.

The whole question of assessing productivity in the service sector is complicated. Let me illustrate this. Students who graduate from the best high schools in the country—private and public—actually know more than students who graduate from the weaker 50 percent of the 1,800 institutions called colleges. That is, a graduate of the Bronx High School of Science, of Andover, or Exeter, or any of the better schools, knows more English, mathematics, and history, and can write better and think better than 50 percent of the average college graduates. In other words, a college diploma alone does not delineate any specific kind of achievement.

We talk about hospitals, but the difference among poor hospitals, medium ones, and good ones can be the difference, literally, between life and death. The difference among professors, even at the same university, can be the difference between stimulating teaching and rote instruction.

How then, in an advanced technology and an advanced economy, can we measure and evaluate the quality of the services that are being produced? All international comparisons about economies are meaningless; for example, the rate of growth of the Russian economy cannot really be compared with that of the Japanese, the American, or the German, unless this qualitative aspect of service is taken into account. We can measure bushels of wheat, square yards of floor space, tons of steel, but when we shift into the service sector, we can no longer make accurate measurements.

A third dimension of work is the investment cycle for training able people.

I will use an extreme case for an example. A study actually made of the members of the American Psychoanalytic Association showed that in general, psychoanalysts complete their training, including military service, at the age of forty and a half. This means that they do not become fully independent practitioners until 40.5 years of age. And since, according to the study, they start to die at forty-one, it is easy to understand why their fees are pretty high!

That seems a fatuous way for society to operate. Regardless of the profession or of how delicate their subject, this makes no sense. To balance off the limitations of human life with the necessity for training in depth, and nevertheless to permit people to begin to practice their profession at a reasonable age, remains the major challenge. I remind you that Pitt was Prime Minister of Great Britain in his early twenties and that Schubert was dead at thirty-two. To contend that there is only one road to excellence—that is, elongation of training—is defeating.

The elongated training problem is really part of the continuing educational problem. Therefore, another facet of the work problem is that, if we shorten education, the question of whether business, or the university, or government will provide for continuing education becomes crucial. The period of initial training cannot be shortened unless provision is made for some continuing training.

Let us consider still another facet of this problem. American industry now has very large numbers of college and university trained personnel committed to a real discipline; these are chemists, physicists, engineers, economists, mathematicians, and even sociologists and a few anthropologists. A basic element of American corporate structure is that an employee is supposed to be faithful and dedicated to his employer. That is the way the structure operates. In turn, the employer has certain responsibilities. However, with professionals this matter is more complicated. Here a double allegiance exists, and the better the professional, the clearer is his allegiance to his profession. Complicated relationships now exist between the chemist or the physicist who is concerned with his science and his peers on the outside. And complications arise because of demands made by and obligations to the company.

Another point is that many skills are not inherent in an individual but relate to a specific environment. An administrator who is a plumber can feel secure if he loses his job; he has his kit of tools and can go out and make a living. One of the phenomena of large-scale corporate life is that a tremendous amount of skill is simply the ability to operate in a particular organization, and this skill is not necessarily transferable. It is knowledge of the history of a particular organization which makes a person useful to it. He knows the distribution of power, the personalities, and the routine. If men do not have a transferable skill that they have built up to a high level of earning power, their relationship with their work can be nervewracking. The question of the nature of human capital accumulated through experience, and the extent to which it is transferable, is an interesting one.

Next we have the question of what these occupational and work trends imply for an institution such as the trade-union movement and for the political structure of the country. The trade unions originally got their strength from blue-collar labor. Their difficulties at the moment stem from the fact that this is, at best, a stable and not a growing sector of the economy. We have powerful unions, but they are not in the sectors of the economy that are growing.

We must not write off completely the potential growth of the trade unions in the white-collar area. We have been living through a transitional period. We have been waiting for the old trade-union leaders to retire—leaders who have no understanding of white-collar workers and no capacity to organize them. But in the years ahead, we will see considerable organization in the white-collar realm. An outstanding development, interestingly enough, is in the governmental sector. We have seen strikes of teachers, nurses, social workers, all successfully concluded from the point of view of the strikers. While the law still frowns on strikes by government personnel, most jurisdictions now permit them to join unions.

White-collar workers will join unions whenever they decide that, on balance, they can do better through collective bargaining than through individual bargaining. As more people go into white-collar work, there will be less mobility for individuals at the lower levels, and it will be more likely that they will use unions to improve their lot. If the leadership of the union movement had not been quite so leaden, it would be making much faster progress today.

From this, other problems develop. Female part-time employees are not easy to organize. And American industry is smart enough to offer many of the unorganized a slightly better deal than the organized. Nevertheless, we must not write off the strength of the unions in the future.

I want to bring up the question of retirement in light of the technological advance of society. If the rapid advance of knowledge of science and technology leads to skill obsolescence, we must anticipate increasing pressure from corporate management for the earlier retirement of employees. There may be a few people aged fifty-five or older who have unique value to an employer; however, most of the other high earners—and there is a connection between earnings and age—can be traded off for youngsters who have been trained in the newer techniques. The constantly renewed supply of people with newer knowledge means that it will become more and more urgent to get rid of older people earlier.

We see at Columbia many people who understand this and are making changes under their own momentum. The School of General Studies has a program under way, financed by the Ford Foundation, to facilitate mid-career changes.

One of the interesting results of a pluralistic economy, which demands earlier retirement for people in certain positions, will be that we may export many of these older people. I know academic institutions where a professor is no longer up-to-date. But he might serve satisfactorily at some Latin American institution or some other foreign college. He will have a better life if he goes abroad. There he will have prestige, self-respect, and a good income, and the university will be rid of an anachronism.

I expect to see a lot more mobility among sectors and, in that sense, there will be earlier retirement. But I do not mean permanent retirement from the labor market; I mean retirement from the first employer. Several hundreds of thousands of officers now coming out of the armed services will soon fit themselves back in the economy and will begin a second career!

We are now in the middle of a major revolution of woman-power. Half of all the women in the United States aged forty-five to fifty-five are at work. If we look back only as far as before World War II, the only married women who worked in the United States were Negro women, widows, and a very few wives of immigrants. A fundamental revolution was brought about by the war. The tradition that said, "Women should stay at home, cook, and care for their children" was changed when we knocked on their doors during World War II and said, "It's your patriotic duty to come to work." And even with our fancy research instruments, we were naive enough to assume that women would go home again when the war was over. Of course, they did not; they continued to work.

Many women work to qualify for Social Security benefits. This is just one of the many reasons that women work, but it is an important one. I think however that in this coming decade women may have more difficulty than in the last decade in getting jobs. In recent years more women than men got jobs. This was a very unusual development, in light of the fact that women account for only one-third of the labor force. That was the result of peculiar demographic developments, and the fact that the men continued their education longer and went into the military. Now, in this coming decade, many more young men than in the last decade will compete for jobs, so the situation may not be quite as easy for women.

It is interesting to consider the South. From 1950 to 1960, there were roughly a million new jobs in manufacturing in the Southeast, the Southwest, and the South. But for all practical purposes, not a single Negro worker in the South got a job in manufacturing in those ten years. The people who got those new jobs were, first, white males, and then white females. There is a pecking order in the South—white males, white females, Negro males and/or Negro females, depending on the kind of work. Now, in the Greensboro triangle, there is, for the first time, a breakthrough of Negro women—not Negro men—into manufacturing employment.

One of the interesting facets of women's working from the employers' point of view is that as more women work, employers may have to begin to think about *family* employment prospects if they want to hire or to relocate somebody. For the first time we have just given a woman a professorship in the Graduate School of Business. Her husband is a physicist, and the question was whether he could be relocated if she moved to New York. They were from the Midwest. Looking ahead, I would say that not only will families be the consumption unit, but if both adults work, we will have to think about the labor market, especially at the skilled levels, as a market for family units.

Our research group has made a study of over 300 educated women over a period of fifteen years. We began to understand that educated women remain attached to the world of work— not that they work all the time, but they are fundamentally attached. The more we educate women, the more likely they are to work. Other things being equal, that is a general rule, and it holds.

The new technology offers us several options, and what we

choose depends on our values. We can have more work; if we want to think up things to do, for instance, we can go to the moon. We do not have to go to the moon, but if we want to, we can. We can have more consumer goods: we can have for instance, three cars per family. Or, we can have more leisure. It is possible to have more work, or more goods, without expanding the work base, or more leisure, or any combination thereof. Historically, we took out about 40 percent of our increasing productivity in leisure. That is, as we became more productive, we wanted more goods, and we took out three-fifths of the increase in productivity in goods and two-fifths in leisure.

I am willing to argue that we will continue to cut down on the hours of work. For some peculiar reason we have had the same number of work hours per week for a couple of years, but the number of work hours per year has decreased. We have, for example, increased vacation time. But I do not doubt that the hours of work per week will go down. The trade unions, for some complicated reason, have not so far wanted to decrease the number of hours of work per week but they will again, especially if unemployment should increase.

Fewer work hours per week will complicate the situation with regard to male employment somewhat, because as the hours of work per week go down, it is easier for women to work. The Catholic Church, which has never wanted married women to work, is aware of that, and does not really desire this adjustment. Here I refer to the many competent Catholics who look at these problems professionally.

The next point is that in the past we used to think about jobs and income as closely related. Economic growth brought more jobs and more income. Those relationships still exist, but they are being loosened. In our kind of society, we can have quite a lot of increase in total output without corresponding growth in employment and income.

There is a group of people, of whom the cyberneticists are

the extremists, who contend that we will not be able to run this society unless we guarantee a high income to insure that people have purchasing power. On the very conservative side, my colleague and friend Milton Friedman says that the way to deal with poverty is through a negative income tax. There is obviously a problem about income in a society where people go to bed hungry and, in the rich United States, it is a scandal that many people do go to bed hungry every night.

I submit that it is not the problem of income that is difficult to solve; it is the problem of work. I do not think we can run a society by simply handing out checks to people without running into new problems. That is, the effective relationship of people to a society is through some effective role, and receiving checks is not an effective role. So, while we may be able to solve the income part of the equation, the work part is more complicated. It is simple enough to send checks to people who live in Watts or Harlem or the Ozarks. We are rich enough to do that. But to give these people a meaningful role—socially and politically—is a different and much more complicated problem.

Some observers are worried that people will not know what to do with their leisure time. People are always worried about an unknown future. I am not worried about this at all because I believe that people can have two lives—a life on the job and a life off the job. And some people I know are having even more. And I think that is desirable. We did a study some years ago called *The American Worker in the Twentieth Century*, in which we found that a large part of the working population, even many years ago, was much more interested in their "life off the job," than that part of it spent on the job. Some of them were concerned with religion; others had different avocations. That is perfectly all right. There is no reason why one should be centered solely on one aspect of life.

I suspect that as the hours of work become shorter, more and more people will become more and more interested in activities off the job, and I think that is good, not bad. An English gentleman with money always wanted freedom to go away on Thursday night and return Tuesday morning, and then he would work in town on Tuesday, Wednesday, and Thursday. Anyone can follow that pattern who has enough education and cultivation and knows how to use his time. I believe that people who say, "The other class won't know how to use their time," are being presumptuous. Of course some people are going to look at television, but after a while they will get bored with it.

The next point, women and their work, is related to the above. Wifehood and motherhood put certain constraints on women and their ability to work. Women are individuals with the right to develop their potential. These are complicated problems of balance. Our book on *Life Styles of Educated Women* is written in terms of options, not in terms of constraints. We felt that women have tremendous options, and we said, "If men are lucky, they will some day have the options that women have." That means that men too would have the option to concentrate on their work only, or on their work and their family; they could work hard at certain periods in their lives, and less hard at others, and so on. A rich society offers more opportunity, with different patterns of options to choose from.

To what extent are we serious about offering opportunities to all the people in the society to develop their potential? We talk about education and we say we are committed to education. Today the amount of money available for education per child in the United States varies widely. Consider the difference between South Carolina, which spends about $250 a year per child in school, and any rich suburb which spends about $1,200 per year. This means that there is about a fivefold differential in expenditures for education between poor and rich communities, between poor and rich states.

If we are really serious about our democratic commitment to equality of opportunity and education—although I do not argue that we have to equalize this exactly—we must attack this fivefold differential, or we are deceiving ourselves. We have a fundamentally unequal society with respect to education. And since education is increasingly the basis for later achievement, we are perpetuating the inability of large parts of the population to get out of their low-income, low-status situation. Unless we get this educational problem under control, we will develop a problem of caste. Lack of education has become a much greater handicap than heretofore when competition was in the place of work itself more than in the educational preparation for work.

In the old days, a peddler, if he were smart, could build up a very good business and, after a while, become a small business-man, and then a larger businessman. The question of whether he went to college was irrelevant. A good example are the great captains of industry who lived at the end of the nineteenth century. None of them had ever seen the inside of a college. As Governor Rockefeller has said, he is the grandson of two high-school dropouts, but he does not recommend this for the present generation.

With regard to the Negro issue, I want to point out that if a minority has been exploited for 350 years and if all kinds of institutions have been constructed to keep it outside of the society, it will take a little while to bring it effectively into the society—especially when most of the majority does not want to do that, or does it only reluctantly.

We have had a democracy in this country for whites only. In my opinion Mr. Myrdal was wrong; he misread American history. We were committed, from our first president to this one, to keeping the Negro on the periphery of the society. Every president of the United States, from George Washington to Mr. Taft, including Mr. Lincoln, wanted to send the Negroes to Africa. They saw no possibility of absorbing the Negro minority in white America.

The basis of slavery in this country was a system of social

control; it was economic exploitation only secondarily. It was inconceivable to the early colonists that these strange, un-Christian, illiterate people could be absorbed into their society. And that was the basis of exclusion. And the free Negro has been the problem. Many of our metropolitan problems are the result of the white population's fleeing from the Negro migration. And although our society is better positioned now than it has been for a while, we face real danger. We have made the first moves to integrate the Negro into our society, but it is just the beginning. This will be *the* challenge of at least the next three decades. My more pessimistic estimate is that only one-fifth of the Negro youngsters ready to work today are going to move into middle-class society. My optimistic estimate is that more than one-half will. These figures refer to youngsters only, and do not include older persons.

The next problem is to know who will pay the price of the change which will result from technology? I have just read a most interesting lecture, which Professor Warner prepared for the recent meeting of the American Economic Association on Technology in the Maritime Industry. One of the questions facing a society which gets richer and more civilized is: who pays the price for the change? Now, obviously, our society has become increasingly interdependent. If people lose their skills and lose their income and lose their seniority, that should be a charge against the employer and/or the larger community. It cannot be a charge against the individual.

We are faced with complicated problems, as the President indicated, in the transformation of the trade-union structure to absorb workers in the critical services. This is another derivative from our changing technology, and our society already has vulnerabilities to which the answers of the old institutions do not fully fit.

I want to end with a double contention: work as a problem is less important than we think and, at the same time, it is more important. It is less important because it is highly desirable that the number of hours in which we are formally committed to a job will decline, as it has declined in the past, for the bulk of the population.

On the other hand, there is no social stability, no survival for democracy, unless all of the groups in a society have an effective relationship to the world of work. The one phenomenon that gave Hitler his momentum was the fact that large numbers of young Germans who had reached working age could find no place in society. There is the danger of alienation when young people see no future for themselves. It makes the problem of work of critical importance. To use an old Talmudic expression, "When you stop working, you're dead."

OPEN DISCUSSION

COMMENT. The areas you have touched on seem to be so complex as to be beyond the hand of human management. If we are going to assume, therefore, that we have to live rationally in an irrational world, we must assume that there is a vast amount of self-correcting inner activity going on in society. Otherwise things are going to fall apart.

GINZBERG. I think that is a very perceptive and important observation. I would reply by saying that one of the great strengths of industrial capitalism in this country has been the market place. It has taken care of a lot of adjustments, and we have resorted to government to take care of certain other things, and between the two, we didn't do badly up to a certain time in our history. The first real shock came with World War I. New orders of problems came with that war and then, of course, there was the major depression of the 1930s, and then World War II, and now the cold war. All kinds of new problems have strained the old market-place type of adjustment.

The markets do continue to do a lot of correcting, as does

the federal government. But I see some problems facing us that will require new adjustments, new kinds of action.

I think the critical problems we face are at the points where we don't understand what's going on. First, I don't think we've yet recognized the overwhelming degree to which human competition has been shifted backwards from adult life into the schools, and almost into the family. We have in this the beginnings of a new status society, because the kind of education you get, the kinds of opportunities you have, will determine what kind of a life you're going to be able to lead. And this has more and more to do with your parents, your social position, and the schools that you go to.

I suppose we're beginning to see this, but not fast enough. We take the school problem seriously enough to recognize that we have to invest very much more for the poor, so that they can really become educated. We know that as between Scarsdale and South Carolina you have a difference, within a "democratic" society, of five times the amount of money spent on school children. That's a serious matter.

The second problem point is the lack of understanding by the American people of the degree of major reform that will be necessary really to assimilate the Negro into American society, now that he's become increasingly metropolitanized. The President of the United States said that Watts is simply a forerunner of the things to come. Who ever heard of a President forecasting further and worse riots? And he did that out of desperation, because he doesn't think people understand what we're living with. And I'm sure they don't.

The Negro and the poor used to be centered on Southern farms, and the rest of the United States didn't have to care about them, because the country moved from New York to San Francisco. We were an East–West country. The South was disjointed from the Union. Now, I would say that the South rejoined the Union, and has exported a lot of its poor whites and poor Negroes. Therefore, the question is one of an entirely different order of magnitude, because there are 1 million Negroes in Chicago. I would say that the likelihood that Chicago will blow up one of these days is very real. It has the highest density of Negro population anywhere, with a somewhat unsatisfactory governmental and employer structure. So I don't think that we know yet what we face in our cities.

QUESTION. May I ask this question? What role, and is it an important one, does the improvement in our communications ability play in this? Is this important, from the standpoint of distributing information to the population generally about these problems? There's an old saying, you know, give people the truth and the truth will make them free.

GINZBERG. I have two views about this. If the pictures on TV of Selma, with that police chief beating up poor Negroes, had not been put on a screen so that all Americans could see, then I think that the rate of change down there would have been still slower, and the people who read this morning's *Times* would never know about it. So, in that sense, the better the communications are, the more national a society you get. That is, you get some sense of identification with people that you don't otherwise see.

But there's another aspect to communications that leaves me very cool, calm, and collected. I'm not impressed with it. There's a lot of internal communication between management and its employees. I once talked to a large conference of corporations at which they estimated that $100 million of that kind of communications is going on. I think most of that's wasted money. I think the notion that management should "explain its position" to employees, through these means, and assume therefore that the employees won't ask for this, that, or the other thing, through their unions or any other way, is very naive. There are many problems in the United States interpreted to be "communications" problems, which I believe to be "conflict" problems, which have entirely different and deeper roots. It's perfectly ridiculous. I think there are some situations where information is useful, and there really are confusions. But in many many places, it's a kind of a namby-pamby approach to say that, "If we only had better communication, everything would be all right." I think that's ridiculous.

QUESTION. If internal communications leave you cold, is it the idea of it, or the way it's being done?

GINZBERG. I think it's manipulatory. I think people are very smart, even poorly educated people, especially when they're being manipulated. So if you really just want to tell them some things that it's of interest for them to know about the company for which they work, that's one thing. But most internal communications that I know of are primarily manipulatory, in their implicit or explicit intent. I therefore think that the people who receive them are very suspicious of why they're being written to. Take one of the large manufacturing companies. They've got an elaborate program in internal communications. It costs the company millions of dollars, but I think every time they send one of those pieces out, either it is not looked at at all, or the employees say, "What the hell is management trying to do to us?"

QUESTION. Would you say the same thing about advertising?

GINZBERG. No, there's a big informational element there about products and so on. It's in the management–labor relationship that so much manipulation goes on. In an open market like advertising you can't do it.

QUESTION. Now that you've brought up labor–management negotiation, let me ask a question about one of the issues we often have to deal with. I've sat across the negotiating table with unions, and we talk about time off to recharge their batteries, and they want sabbatical leave, and all that. They don't say what they're going to do with their time off, other than just get paid for it. What do you suggest that people do to make these

leaves, of whatever duration they are, be even better for the job
that they had when they went away?

GINZBERG. That's a good question, in terms of kinds of people and/or alternative programs. Let me quickly say a few
things. We have states in the United States which, by constitution, prevent any public school money from going to anybody
above the age of twenty-one. That's except for the state university, which is a separate line on the budget. That means that we
have a conception that education is limited to children and
young people. It's quite clear now that that's a misconception,
and doesn't make any sense. In an affluent community like Santa
Barbara, you will see a most fantastic, elaborate, adult educational program, for technical skills, for avocational reasons, for
retirement preparation, and just for cultural learning. A great
program. I have been a consultant to the U.S. Commission of
Education, on the rebuilding of its adult education program. It's
lagged. So you will begin to see the federal government trying
to nudge the states now to modernize adult education. That's
one way to use this leisure.

There are some unions, including the Electrical Workers
Union and the Plumbers, that have been running very expensive
retraining programs for their own journeymen. My recollection
is that the Electrical Workers spend about a million dollars a
year, which is a lot of money for a union, just to upgrade their
own journeymen so they can deal with new technology, because
an ordinary electrician can't handle the new stuff unless he's re-
trained.

I would argue that, for scientific personnel, single-shot programs are not as good as the remission of part of their time on a
continuing basis. There are many ways of doing that. In the
atomic energy organization in England I believe their recruitment terms are that you work 50 percent on in-house research,
and 50 percent on that part of the research that has special mean-

ing for your own development—I mean, what you're interested in. That's another way of doing it. I've argued with Du Pont for years, saying to them, why don't you let as many of your scientific people as want to become adjunct professors get off in the afternoon. What do you care when they're off? The notion of having to hold them in the plant, from 9 to 5, is unrealistic.

QUESTION. You suggested that people will have more leisure time in the future. Working hours presumably for blue-collar workers and a certain number of white-collar workers will be reduced by about 40 percent. But will managers, professors, thinkers, administrators, and so forth have their working hours diminished in the future? Is it possible? If it is not possible, people will continue to work a 60- or 70-hour week in these jobs whereas the rest of the population will be working, let's say for argument's sake, a 25-hour week. Will there be a growing gap between these two segments of the population?

GINZBERG. I think that we shouldn't exaggerate how much management works. We used to have a six-day week. As far as I know, most management now doesn't work a six-day week. It works a five-day week, at least officially, when it goes to the offices. The overriding general pattern of work in society affects everybody in the society, although somewhat differently. There will always be a few people who work, even if they don't go to the office. Or they get a key to open up the place. But in general, I would say that there tends to be an adaptation, made in different ways, of everybody who works, as the general hours of work go down, and particularly as the vacation periods increase. There's such a mix-up in this country anyhow, between folks' work and certain associated leisure time activities. So much business is done at the golf club, during social activities at night, or at lunch. This is a peculiar pattern that we've developed in the United States. It's not so true always in Europe, where there's a sharper delineation between a man's personal life and his work life. We've got it all mixed up, so that in industry they take the wives of the executives to Puerto Rico for a training program, etc.

I would say that in general, by and large, the American public will be "able to absorb" the additional leisure. And it's going to come slowly. It doesn't come in any spectacular units. The work-week hours go down, the number of vacation days goes up—that's the way it moves.

I'll tell you about a piece of research done many years ago on movie projectionists. They were working 28 hours a week— four days on, three days off, three days on, four days off. I started with your "puritan predilection," expecting to find all kinds of neurosis, pathology, difficulty, family tensions, and so on, with men around the family that much. I got the shock of my life. I found that these quite simple people with not very high earnings, not very well educated, didn't have much problem with their leisure time. If they had younger kids, they used to spend some time taking them out in the park. Some of them like to play chess, and went to the union headquarters and played chess. Some of them like to read, so they went to the library and read. Most of them helped their wives a little bit with shopping, etc. Some of them painted pretty well, so they spent a little bit of money and got some materials and started to make models and so on. A few of them moonlighted, but very few. That's also a mistaken notion, that everybody's going to turn his time into moonlighting. All I can say is, by and large, they felt no special pain, and they did not have a lot of money to spend. My general view is that it's somewhat presumptuous to assume that people are so poorly educated and so lacking in potential that they will not be able to do something reasonable with the time at their disposal.

Change and the Less-Developed Countries

EVERETT M. KASSALOW

WHAT ARE THE PROSPECTS for change—technological, economic, social, and political, for they are all interrelated in development—in the young countries? [1] Can we make generalizations about so complex a process in so many and varied countries?

It is possible to make generalizations about the process of change in the developing world, provided both the lecturer and the listener constantly bear in mind that the experience everywhere is somewhat different. The generalizations have limits, and there are exceptions to them and exceptions to the exceptions. But like all efforts at scientific explanation, these generalizations must be attempted, even while recognizing the limits. There is really no other path to understanding most complex phenomena save by some degree of generalization. To refuse any effort at generalization is to leave us wallowing forever in minute details.

It seems wise, however, to begin with some of the obvious limits and qualifications about what can be said generally about

EVERETT M. KASSALOW IS PROFESSOR OF ECONOMICS AT THE UNIVERSITY OF WISCONSIN.

[1] The literature in this field is, of course, enormous. In addition to some of those sources specifically cited, I have found particularly helpful, in recent years, the work of Simon Kuznets, W. E. Moore, Theodore Schultz, and Alexander Gerschenkron.

change and the developing countries. First, among the many countries under discussion, size alone makes for important differences. As one searches for generalizations he must be aware, for example, that there are almost as many people in India as in all of Africa and Latin America combined. India is nearly fifty times as large as Kenya, and Kenya is not a small country by African standards. This difference in scale makes the achievement of political consensus, the possibilities of a market, the variety of resources, and other phases of development, a more complex matter in India than in many of the other new nations. Most of the developing countries, however, regardless of size, do not themselves lack in those complicating factors such as regional, language and tribal, or communal diversity which are part of India's burden.

Second, political backgrounds and conditions vary enormously. Political independence is, typically, one hundred years old and more in most of Latin America, but only one or two decades old in most of Africa and Asia.

Where, as in parts of Africa and Asia, national boundaries are still not precisely defined and remain largely unknown to the bulk of the citizenry, where any sense of nation is, at best, only a few years of age, the impact of economic change often has a different force about it as compared, again, to most of Latin America. The structure or institutions which "contain" or guide economic change are, in a broad sense, political, in any given country. When these political institutions vary so greatly, one must expect great differences in the impact of economic change.

Yet, there are many important, common factors about the impact of economic and technological change upon the less-developed countries, which permit us to generalize. Foremost of these, perhaps, is that within the past few decades the desire for industrialization has become univeral. There may be isolated tribes or islands that are not interested in industrialization, but for practical purposes all the nations of the world, and particularly their leaders, are now deeply committed to industrialization. The day when anthropologists or sociologists could criticize the economist on the ground that the latter was extrapolating his own biases when he projected economic values onto countries where the market and the drive for industrialization were not operative has passed, at least so far as the drive for industrialization is concerned.

TRADITIONALISM AS A RESISTANT FORCE

The goal of industrialization, then, has become a universal one as far as countries, East and West, are concerned; but this goal is often understood and accepted only by the leaders and a relatively small portion of the citizenry. The goal is just as often unknown to the general populace in many of the non-Western countries. Unlike the case, say, of the United States, where the great mass of people are well caught up in what one writer has called the desire and struggle for "enrichment," in many of the new countries such a desire remains unknown among average citizens. This, in itself, constitutes a barrier to change and technological advance.

A young Indian sociologist, Kusim Nair, only a few years ago sat down in a series of villages to probe into what appeared to be the factors delaying industrial progress in her own country. Despite the conditions of extreme poverty in the countryside, she often found a curious indifference or lack of understanding, or lack of aspiration, for any change in economic conditions. As she tells one of her stories, in the State of Madras she gathered a group of Harijans (outcasts living in mud and straw houses, in abject poverty) about her:

> "If the government were to offer to give you as much land as you want, absolutely free of charge, how much would you ask for?" I ask them. . . . "You would like to have land of your own, wouldn't you?"
>
> "Yes." Many heads nod.

"Then how much?"

Baemu is the first to speak. He is an old man. He had never possessed any land. There are five members in his family. But he wants only one and one-third acres. He is precise. Even from that, he says after some mental calculation, he would be prepared to share 50 percent of the produce with the Mirasdar. Rangarajan is middle-aged, tall and slim. He also has five in his family to feed, and two acres would suffice. Mamickam, with six in the family, already has three acres on lease. If he could, he would like to add to it two more acres. Srininasan looks an artist. He wears a beard and shoulder length hair. No, he is a mere labourer, has four in the family, wants just two acres, and is prepared to share the produce on a 50-50 basis. . . . And finally it is the same story with another young hopeful—Velayuthaw. There are nine in the family, and at present they are cultivating three acres. Yet he asks for only three acres on a 60-40 basis. . . .

Scarcely able to believe their answers, Kusim Nair asks, "Are you sure you would not like to have more?" But they are quite sure, leading her to conclude:

The sun is settling on a roseate rice-girdled horizon. Mental horizons are similarly circumscribed. . . . Their minds and aspirations are calculated solely on the basis of the family's consumption requirements of rice at two meals a day, one of which is cold and left over from the previous night. For the time being, at least, they do not want more.[2]

This limited horizon, a lack of relationship to any of the goals of industrialism or modernism, is a significant barrier to change and economic progress. Even where workers may be caught up in urban, modern sector jobs, frequently they will leave without notice, pulled back to the land of their fathers and forefathers. Often, it is true, unbearable living conditions make city jobs intolerable, but often, too, it is the pull of the traditional way of life which brings them back to the village.

Industrialization and urbanism are so implanted in our lives that we forget how revolutionary a process it is, how much it

[2] Kusim Nair, *Blossoms in the Dust* (New York, Praeger, 1962), 30–31.

denies the traditional values that every society, even including our own, has tended to hold dear.

INDUSTRIAL CHANGE: THE CLASH WITH TRADITIONAL SOCIAL VALUES

The manner in which these traditional values are shattered by economic modernism has been strikingly caught by the French scholar Bertrand de Jouvenal, who observes: "The conditions necessary for general enrichment . . . are of a Draconian severity. In the forefront of them stands the mobility of labor." In contrast to the past, in the new order, "a man must be prepared to change his way of working, his occupation, and his place of existence, to vary his techniques, to do something else, to live somewhere else." Men must be prepared to change their jobs and living places constantly, "For that is an imperative of productivity."

The result, as De Jouvenal notes, "is a reversal of every social value." The idea that a man must have roots, hold dear the land of his fathers, the place which holds the memories of his childhood and his youth, now are only obstacles to making man a good producer. His job, too, the craft he may have learned in his youth, under many circumstances, "must go out the window, if technological progress is to be served, if a man is to become a true citizen of the city of productivity." [3] While De Jouvenal's prose is quite dramatic, he does convey a sense of what the commitment to technological progress involves. In the light of these enormous consequences, it is easier to understand the forces resisting economic change.

Our own society is full of examples of occupational groups resisting change. Among labor unions, the locomotive firemen and flight engineers who clung to their jobs, and sought to preserve them after society had decided they were technologically

[3] Bertrand de Jouvenal, in Edward Shills, ed., *The Problems of Afro-Asian New States*, Encounter Pamphlet No. 1, 1962, pp. 10–11.

obsolete, are good examples. The several million farmers who cling to the land, even though their economic contributions are almost nil, are another case. In this latter instance, a nostalgia for a past way of life even leads society, or Congress, to pass to some of these economically obsolete farmers enough of a subsidy to keep them alive. Similarly, during the first years of computerization, observers here and abroad frequently note that different levels of management often resist its introduction for fear of being eliminated, or losing power.

It is not occupation or location alone which economic change overruns, as it virtually floods out the older civilization. The process of modernization in the Western world, including the rise of capitalism and the national state, also encompassed a revolt against the prevailing religion and god of the medieval church. Traditional religions and gods are not likely to fare much better as the new nations embrace modernism.

On the more secular side, contemporary sociologists have concluded that what is popularly called the generation gap is primarily a by-product of a technological civilization, dedicated almost obsessively to change and progress. What was good and true for one generation obviously will not and can not hold for another generation, in a society in constant change.

Finally, on this general problem of change, we should note that a process which took three or four centuries in the West is being collapsed, hopefully, into three or four decades in the young countries. All the shocks, tensions, and neuroses which go with change, then, are likely to be multiplied by the rapidity of the effort.

WHAT NEW SOCIETIES CAN LEARN FROM WESTERN EXPERIENCE

How closely will economic and social development, what for summary purposes we can call the path to modernism, resemble Western experience? Instinctively most of us feel that ours has been a universal experience, and that a new country's development will repeat ours. Karl Marx once commented that if the less-developed countries want to see their future, they have only to look at the developed countries, and read their history, and they'll understand what they must go through.

Will the new societies simply, basically repeat Western industrial and political evolution? Will there be the dreary "satanic mills" of the late eighteenth and early nineteenth centuries —with primitive capital accumulation, child labor, twelve-to-fourteen-hour work days, and the rest?

This seems unlikely. A variety of forces will make the evolution of a new country critically different from that of the West, be it the capitalist West or the Soviet West. (In a sense, communism is a later path to industrialism—only a derivative of capitalism, which was the *fundamental* break with all prior civilizations, and which one way or another remained within the bounds of traditionalism.)

CLIMATE AS A FACTOR IN NEW COUNTRY CHANGE

The very climate of most of the new societies sets the problem differently. Until now, nations successful at modernization have all been in the temperate to cool zones. The new nations of Asia and Africa (and, to an important extent, Latin America) are overwhelmingly tropical or semi-tropical. The very notion of and commitment to work, so all-central in Western development, may never assume such full dimension in Asia and Africa. The need and the desire to return to the countryside for rest, recreation, and retreat may continue to be greater, even after some substantial industrialization comes, in tropical and semi-tropical areas. Even in southern Europe the way of life has been something of a holdout against full modernism. Why should we not expect something even more different from the typical Western pattern in those many new countries in the warmer climates!

THE POLITICAL STARTING POINT OF THE NEW COUNTRIES

It is useful to compare today's new societies with those of the Western world, as the latter stood in the early decades of the nineteenth century, poised on the brink of modern economic development, with their populations still overwhelmingly rural, just as the countries of Asia, Africa, and, to some extent, Latin America are today.

One major difference between most of the nineteenth-century Western societies and those of Asia and Africa today is the relative political development. For countries like France, England, the United States, Switzerland, and several others, national political consensus had been largely achieved, and the formulations of the modern, rational state had already been laid down.

The new nations of Asia and Africa are just that—*new*. They have barely won their independence. They lack stable political structures and consensus among groups or regions. The notion of a national market, the acceptance of a national government and a fixed territory, barely exist or do not exist at all in most of Asia or Africa and parts of Latin America.

The leaders of the new nations must simultaneously build effective political and industrial structures. What the Western countries did over centuries, the building of the modern political country and state, must be done almost immediately in the new societies. At the same time they must begin the drive for industrialization.

Moreover, the economic structure of the Western world was, to a very important extent, built by private interests, the capitalist–bourgeoisie class. In the new countries such a class is largely absent, and the leaders of the state have the dual task of building the modern economic and political structures. The strain on all is inevitably enormous, and often insupportable.

POLITICAL DEVELOPMENT MAY TAKE PRECEDENCE

It should not be surprising to find that very often it is the political rather than the economic thrust that gets the prime concentration. The leadership of the young country recognizes that unless it can hold power, most everything else is of little consequence. Political judgment colors all other decisions. Who should be appointed to run the railroads, the new dock construction, the new steel plant? The best technician or administrator? or someone who is certain to be loyal to the new, still barely understood or accepted national government? If the technician is passionately loyal politically, fine; if not, the leader feels he has little choice but to name someone perhaps less proficient technically, but whose loyalty is unquestioned, as the appointee will head a large concentration of wealth, jobs, and prestige. It is not too easy to fault the country's leadership, here; for if there is no political stability, economic development can hardly proceed satisfactorily. On the other hand, if the decisions are made totally on a political basis, with no regard to technical need or efficiency, economic chaos can result.

The problem is beautifully conveyed in an Independence Day speech of Indonesia's Sukarno in 1964, as he struggled with nation-building, in his own terms. Addressing the multitude, he declared that: "The prime minister of a foreign country said to me one day, 'How can your country survive if you have no heavy industry?'" For Sukarno, this question only showed "how stupid this prime minister was. He thought that the life of a nation depends on the technical level of the country and its industry." Untrue, cries Sukarno to the Indonesian masses: "No sir. The life of a nation depends on its national consciousness, and the life of a revolution depends on its revolutionary consciousness. Not on technology! Not on industry! Not on factories, or aeroplanes or asphalted roads." [4]

[4] Quoted in Herbert Luethy, "Indonesia Confronted," *Encounter* (December, 1965).

With the advantage of hindsight we can perhaps say that in the longer run Sukarno is wrong and a nation cannot proceed into modernism and stability without technological, as well as political dynamism. But it is well to recall that despite abysmal economic management Sukarno survived for twenty years on the basis of political charisma and manipulation, and he is still not dead politically.

ECONOMICS AND POPULATION: DIFFERENCES IN
STARTING POINTS

Simon Kuzmets has pointed out that the economic formulations of most of the new states in the twentieth century have been different from those of Western countries at comparable launching points in the nineteenth century. Most of the new societies begin their industrialization drives with an annual per capita income of $70 to $100. Western countries began on bases ranging from $100 to $300 in the early nineteenth century. Even though nineteenth-century European countries were, as the new societies today are, agricultural, it appears that even agricultural productivity was higher in the West, at a comparable earlier state. Again, today the country of $100 per capita income exists in a world where some states can boast of $1,000, $2,000, even $3,000 per capita levels. The tension of rising expectations set off by these contrasts are now commonly understood, but they are no less acute. The nineteenth-century European (and American) industrialists had no similar competitors with high incomes.

Even more acute is the population problem. At the take-off time in the West, the early nineteenth century, there was no really large country. Britain, the leader, had as many as 35 million people, but it was already firmly on the road to industrialization.

Examine the developing countries today. They include India, with around 475 million people; Pakistan and Indonesia, with around 100 million each (not to mention Japan, which can no longer be classified as "developing"); Brazil, with around 75 million, Nigeria, with 55 million; and China, with somewhere over 700 million.

The capital needs of such countries are much greater than what was earlier required in the West, if industrialization is to go forward. How much more investment must be made year after year, in roads, schools, hospitals, public buildings, and dwellings just to accommodate the population.

Even more difficult than sheer population size, from a development viewpoint, is the rate of population increase. In the first half of the nineteenth-century Western European country populations were showing a net increase of 1 percent a year. (The United States with immigration, at times, and Britain, already well industrializing, were exceptions.) Typically, today, the developing countries have rates of net population increases of from 2.4 or 2.5 to 3.0 percent annually. These effects on growth, and especially on living standards, of these population increase rates can be catastrophic.

Take India and Mexico, two countries with a fair industrial base and a good gross economic growth rate (at least until this past year or so). Ansley Coale has shown how efficient the states can be in getting the population rate under control in these two countries, where the net population increase is about 3 percent in Mexico, and around 2.5 percent in India.

According to Coale's projection of gross national product, consumption, investment, and related matters, if India and Mexico "modernize" their birth rates, in effect cut them in half in the next twenty-five years, the longer run effects can be staggering. They will also be staggering if these countries fail to cut these high birth rates.

If the rate is cut, as suggested, after fifty years the country will enjoy an average per capita income which will be higher by 86 percent than would be the case if the rate of population increase is not checked. If this check to population is accomplished, after one hundred years the difference in per capita income could be 234 percent, and after 150 years, nearly 500 percent. The per capita income rate goes up sharply after fifty years, because the childbearers of tomorrow and for the next twenty-five years are already born, or will be born in the next twenty-five years when the birth rate is just being brought down, under these assumptions.

In terms of change and economic development, then, the population issue is as critical as any other, and more critical than most. The new countries too, unlike the European countries of the nineteenth century, also lack an America, a Canada, or an Australia, and to some extent a Latin America, all of which were utilized by Europeans to export some of their growing populations. There are no land frontiers of any major size open to most of the new countries.

The general consequence is that population tends to outrun or nearly outrun economic advance. As Gerschenkron concludes: "Industrial progress is arduous and expensive; medical progress is cheaper and easier of accomplishment." Moreover, the medical progress often precedes the industrial and produces a "formidable overpopulation," and "industrial revolutions may be defeated by Malthusian counterrevolutions." [5]

WILL NEW SOCIETIES FOLLOW WESTERN
INDUSTRIALIZATION?

With these significantly different institutional background differences, it does not seem likely that most of the new societies will follow a relatively slow, "easy" evolution into industrialization—such a path as most Western nations pursued in the eighteenth, nineteenth, and early twentieth centuries.

It is likely, for example, that more can be learned about the

[5] Alexander Gerschenkron, *Economic Backwardness in Historical Perspective* (New York, Praeger, 1962), 28.

process of industrialization in the new societies from a study of "what happened" in Japan and/or the Soviet Union, relatively late developers, than a similar review of earlier developers like the United States or Great Britain. The experience of Japan and the Soviet Union suggests, for example, that to overcome backwardness more rapidly, under pressure of today's political and social tensions, there is apt to be a greater mobilization of the society, a more conscious and diverted effort.

In their desire to top the enormous accumulation of scientific and technological know-how now available, the new societies will assume more direct control of the industrialization effort. This control will almost certainly be state control, by the nature of things. The absence of a bourgeois middle class (hardened as it was in centuries of commercial and industrial development in the West) often leaves the state with little choice but to play a major role in accumulating capital, making investment decisions, and generally guiding development.

The very size of the effort in many industries—you do not, for instance, start with a small iron and steel plant but almost necessarily jump into a new modern steel complex—puts the burden on the state. Who else can mobilize the capital and put the society in motion to build such a new industrial empire? A much larger degree of state control has already emerged in practically all of the newly developing countries. It should be added, as a kind of footnote, that this greatly enlarged state role in economic direction is not always necessarily better than a great reliance on more private effort might produce. Malaysia and Formosa, which have made relatively greater use of private forces, show more rapid growth than most of the more fully state-directed economics of their neighbors. Even this question of a state control is, however, only a relative matter. The hatred of the colonial past, and its identification with imperialist capitalism, makes the latter unpopular nearly everywhere in the new societies. With but a few exceptions they all pledge themselves to "socialism," although the word tends to take in so many shades of meaning, it becomes more and more difficult to find real meaning in it, save, perhaps in its negative, suspicious attitude toward Western capitalism. Many of those very countries who proclaim their faith in socialism, nevertheless, seem more than willing to accept, and indeed often seek, foreign private investment. The net result is, and will be, a hodgepodge of new social and institutional forms; but nearly everywhere the state has the crucial, directing role.

The magnitude of the effort and the desire for acceleration of development in the new societies also seems to carry with it the need for some sort of a development ideology. The process of economic growth was more or less spontaneous, without conscious plan, in the United States and Britain. In the new societies, faced with great visible disparities in living standards between their own and Western countries, and burning with the conscious determination to mobilize the population for an accelerated effort, some higher appeal by the state seems in order to galvanize a still tradition-bound population.

THE PROBLEM OF FEUDAL REMNANTS

Since this effort is a mobilized one and aims at relatively accelerated development, where there are still major unresolved social questions, the process can easily explode into revolution. It was, after all, the land question, as well as war, which gave the Bolsheviks their great opening as Russia floundered about in her effort to make the transitions to modernism. The land–peasant question seems to have been an important factor in the triumph of the Chinese Communist Party. It is hard to estimate what might have been the political consequences of a link-up between the peasantry of Japan and the Marxist-tending working classes in that country, if extensive land reform had not been accomplished, under the level of the American occupation, in the immediate post-World War II period.

In contrast, feudalism was well liquidated, at least in its fundamental economic forms, in Western Europe before the real economic take-off began. Even then, feudalism's cultural, legal, and social heritage helped produce working-class movements and parties dedicated to revolution, albeit not necessarily violence, nearly everywhere in late nineteenth- and early twentieth-century Europe. It was only in the course of many decades that this outer skin of revolution gradually peeled off and socialists became reformists in Europe.

It is easier, in the light of this piece of comparative development history, to understand the revolutionary ferment in Latin America. Most of the nations of that continent approach the jump into modernism with a land-holding and class structure which is more feudal than modern. We can hardly have seen the end of Latin American revolutions.

The picture is, of course, not all glowing. To return to the theme of state leadership, the fact that the state leads the process can also provide many advantages, as well. The conscious accumulation and application of already proven and demonstrated industrial techniques can provide breakthroughs which were virtually impossible in nineteenth-century Western development. Japan is, of course, a shining example, industrially, of what can happen when the process goes well, and the potentialities are tapped. Great advantage has been taken, in Japan, of already proven industrial techniques, and of the available pools of human skill. (Experts in science and industry were deliberately imported into Japan in the late nineteenth century, and afterwards, to help train Japanese personnel in the new industrial tasks and sciences.)

MANAGEMENT PROBLEMS

It is easy, in reviewing Western industrial experience, to take management as given. We can overlook the fact that management, the business class, Marx's bourgeoisie, was the spear-

head of Western industrial development. It was the product of
the long, slow (as it now appears) rise of towns with, first, the
itinerant merchant, then the large mercantile family houses, the
banking combines, early industrial entrepreneurs, and the late
owners of power-driven factories. All of these managerial-
owner types thrown up in the course of different stages of a
long development, were, so to speak, the unconscious guardians
of the industrial processes. They were hardened in what was
generally a nonbusiness-oriented environment, at least until the
nineteenth century. A premium was placed upon their skills as
independent organizers and innovators.

It is easy enough today, in the midst of giant, anonymously
run corporations to forget how individual and special was the
business class which pioneered the economic beginnings and
traditions of modern Western capitalism. The rise of the businessman, the manager–entrepreneur, may seem to have been an
automatic, evolutionary matter, but its importance and complex-
ity are not to be minimized, if we are to understand the management gap in the new societies today.

Clearly the new societies will not experience the same slow
evolution which has produced today's Western managers and
the managerial tradition. The new countries start from a very
primitive base, and jump, in particular cases, to large modern
structures, various aluminum reduction enterprises, or totally
new integrated steel complexes. In building, usually with the
help of Western firms, the new societies frequently find it is
relatively easier to construct the modern plant than to manage it
and run it effectively.

And for many reasons foreign management is, at best,
acceptable for only a brief period of time, if at all. Under colo-
nialism it was natural enough to import foreign management
with the foreign firm. In the new air of independence and nationalism, the new countries want to manage and run their own

new firms. But there just aren't enough managers in the private sectors, let alone for the great new public enterprises.

This problem of finding managers, good managers, for the many new public sector enterprises is common almost everywhere in the developing world of Asia and Africa. It is one which our aid officials, for example, have scarcely faced. Whether we like it or not, there will be large sectors of public enterprise, but it is my impression that U.S. aid officials have scarcely thought seriously about the forms of that enterprise.

Public enterprise, or experience in such countries as Britain and France demonstrates, can be reasonably efficient just as it can be inefficient and wasteful. The forms of public enterprise need careful study since what we recommend, what the new societies adopt, can help make or break managerial effectiveness. In some instances, be it coal or electricity or railroads, for example, both good and bad forms of public enterprise, with reasonable standards or measures of performance in the good cases, are to be found. There is good and bad organizational know-how and technique that can be exported to the new societies, along with the new railroad, hydroelectric plant, or steel mill.

This problem is especially difficult for American officials with their almost built-in social and cultural hostility to public enterprise—let alone the hostility of the Congress that controls aid appropriations. On the other hand, I think there is much to be learned from British and French experience, where public ownership and operation has a long history.

The problems of management, and the sources for managers, of course, vary from country to country. I daresay the problems, as development goes forward, may be most severe in Africa. Here, for so long in important enterprises, there was only foreign management, white management (sometimes Asian in smaller enterprises), and almost no native management cadres were trained.

THE INDUSTRIAL INTEGRATION OF THE AFRICAN
WORKER

Not only native managers, but even native workers have
often been less integrated into the industrial process in Africa.
Rated as an inferior racial being, the African had a sense of limit
and inferiority almost bred into him. Under these circumstances
industrial work itself often became a demeaning experience.

The gap between management and the worker is well illus-
trated in a story told of the experience of one South African
firm. This was a new company, determined to make a good start
in labor relations. It hired young workers, sixteen or seventeen
years of age, and paid good wages. The results were very prom-
ising to begin with, although the firm realized it involved some
rules, since at about the age of eighteen the workers went back
to their tribes for puberty–circumcision rites.

The management was assured, however, that it was worth
making the investment in these people, who, it was presumed,
would return at the end of the rites. None of the workers ever
came back. They had appeared satisfied, produced well, were
well paid, but they didn't return.

This management had failed to comprehend that it was
symbolically impossible for these men to return to the same jobs
(or establishment) they held before puberty rites. Such jobs
were associated with childhood, beneath their new status as full
members of their communities.

Eventually the company shifted its hiring requirement to
eighteen years of age and above, and the problem was overcome.
But how many other cultural–communication barriers to effective change and development exist in Africa (and in other new
societies) one can only dimly perceive. In so many instances
they continue to prevent that true bulwark of factors in the
enterprise which is so crucial to improved production and better
human relations.

Our own Western experience has scarcely equipped us to understand many of the forces resistant to economic and technological change and progress in the less-developed countries.

CHANGE AND THE IMPORTANCE OF HUMAN RESOURCES

This turns us to the final theme. Management, after all, is only one part of the human resource question. It is the whole of human resources that must be transformed and mobilized if rapid change is to be the order of the day in the new countries.

Theodore Schultz suggests that the cycle usually runs as follows: first the new country builds a large steel plant, and then believes it has industrialization; it discovers that the plant lacks managers, and begins to train high-level manpower. Then it is realized that things cannot be run effectively without trained and committed workers. After all this, officials of the new country discover that agriculture has lagged, and that development is still not effective. The need for human resources, and skilled human resources, in agriculture must then also be met, if development is to move forward favorably.[6]

Too often our vision of development, whether capitalist or Marxist, has tended to be a simple one of capital accumulation. Increase the capital–labor ratio, and more tractors, more machines, more horsepower and economic growth will proceed. Our unusually great success in the Marshall Plan, when our exports of capital aid sparked a great European recovery, helped to mislead us. The process appeared simple as we comfortably surveyed the successive economic miracles in Germany, France, and Italy.

When we came to try a somewhat similar formula in a few of the less-developed countries in the late 1950s, results were much less encouraging. This experience, plus a restudy of the

[6] Theodore Schultz, *Transforming Traditional Agriculture* (New Haven, Yale University Press, 1964), p. 198.

growth process as it has actually occurred in the last sixty or
seventy years in the United States and Britain has revealed to a
number of economists that while the role of capital is important
to economic development and growth, it is by no means so all-
central or exclusively the strategic factor.

In the United States and Britain it is now conceded, conventional inputs of capital, machinery, plant, or equipment will
account for no more than one-fourth to one-half the economic
growth rate. The remainder of the growth seems to be due to
that intangible factor, the improvement in human resources.
And by human resources is included knowledge, technique,
health, education, morale, and all those other difficult-to-
measure "investments."

For investments they truly are, economically speaking. Although we cannot precisely quantify the relationship between a
society's investment in human health or education, the correla-
tion between these investments and those countries which show
a more economic growth rate are quite significant.

The German or French miracles, for example, are easier to
explain when one recalls that centuries of investment in education and infra-structure had created a sense of order, a national
state apparatus, human skills, and discipline of purpose. The
same billions of capital invested in countries lacking these qualities would produce almost no real change, we have come to
learn since the Marshall Plan.

The so-called Japanese miracle of modernization seems,
also, to be primarily a product of superior human inputs and
outputs. Schultz, for example, suggests a comparison of post-
World War II agriculture in India, China, and Japan. Of these
three nations, Japan had the greatest population density, the
worst land-water ratio, and the poorest land. Yet, since the end
of World War II, Japan has made itself nearly self-sufficient in
food, whereas agricultural production continues to lag in India

and China. The difference, Schultz is convinced, lies in the human factor.

An educated Japanese population was in a position to take advantage of the enormous know-how and experience of scientific agriculture that had accumulated in the United States and elsewhere. The fertilizer and improved methods were crucial, but understanding them and being able to understand how to use them, as a result of the prior "modernization" of the Japanese people, was even more important. The same technician could not be employed in nearly as effective a fashion by the still backward, largely uneducated, and tradition-bound agricultural populations in most other countries.

As the new countries face the problem of change, and as we strive to help them in bringing it about, in the years to come the appreciation of the human factor in all its aspects must count for more than it has until now. The strategic investment in human resources, when combined with conventional capital and opportunities for trade (in all of which the advanced countries must be prepared to help), can produce a more successful growth pattern in many new countries.

It is realistic to expect, however, that this path to development is likely to be a rugged one. Western society still bears the scars of past development in the forms of pockets of poverty, slums, and important groups of people who have been stranded. With so much ground to make up politically, socially, and economically, the new societies cannot but find the process of change a hard road.

OPEN DISCUSSION

QUESTION. Would you say something about how development of these new nations will, directly or indirectly, and in the near future, influence our own technological development?

KASSALOW. If the process proceeds rationally, and we see some evidence that it will, they will be moving into areas of production which we will have to vacate—soft goods, light manufacturing, and so forth. Will we move out of these areas gracefully, meekly, changing our labor force with sufficient rapidity? I'm just not sure. South Korea, Formosa, and other countries are beginning to emerge as serious textile producers. Are they going to find enough of a market in the Western world, including the United States? Will we be willing to let our textile industry decline? It is probably logical that we should. We would be better off with our people in electronics production, computer production, and so forth. Whether we will respond this way to the needs of these countries as they develop, I'm just not certain. You know, we tend to be protective about our industries, both labor and management. We have regarded protectionism as a necessity for smooth economic development. But I think we're in a better position to adjust our industrial range to world needs today than we were about ten years ago, with our higher rate of economic development, and with our better manpower policies. But I don't know whether we will really pursue that logic, and let our textile industry run down another 50 percent in the next ten or fifteen years.

QUESTION. You have emphasized that the objective in these countries is always industrial development. But their industrial development is competing with their political development. I have two questions on this point. First, do you believe that their progress will be made more rapidly through industrial development, or perhaps through agricultural or home development? Second, is their principal developmental need for education, for capital, or for management?

KASSALOW. They need modern industrial development, which includes improved agricultural methods. If they stay totally or very heavily dependent upon agricultural development and agricultural production, they will be in a vulnerable situation in the world market and will probably find the terms of trade turning against them over and over again. They cannot accumulate much capital that way to go ahead toward industrialization. Sooner or later they have to turn to industrialization.

Now, this doesn't mean that they can neglect agriculture, because the populations are large and they cannot afford to import much of their food. But the process is usually a simultaneous one. Look, for example, at our agricultural growth in the United States when we were developing as an industrial nation. Scientific feedback is very great. Productivity tends to spread from one sector to another.

Now, on the question of education, capital, and management—it's hard to separate these things out. Those who've been trying to study the educational and manpower needs of the new and developing nations have come to a cautious conclusion that the new nations probably ought to concentrate more of their resources than they have tended to on high-level manpower, rather than upon primary education. That's understandable; from a strictly technological and developmental point of view it may well be right. I suspect, however, that because of the needs of politics and citizenship no country can afford to neglect primary education. If there is to be a hope for democracy in these nations, I for one think that an emphasis upon primary education, with all of the problems it creates, is a vital necessity. It means you have to neglect higher management education or skilled education to some extent. We create expectations on the part of people, if we educate them, whereas if we kept them or left them in ignorance perhaps they would be more placid. But I cannot see how we can have much or any hope of democratic development unless we bring the majority into the whole process, and primary education counts, I think, in this respect.

As for capital, their needs are almost limitless. Their ability to absorb is somewhat limited by the human factors. We have been finding in a number of cases that we export capital to some of these countries and they can't use it too constructively, not all of it, because they don't have the manpower skills, they don't have the managerial skills. I suspect, as they get their economies and societies in more rational order, they will be able to accept and employ this capital to a much higher degree. Management goes with capital, I suppose. It goes with the building of the substructure of roads, dams, highways, and so forth, as well as with private enterprise.

The real tragedy is that with the populations we're talking about, the needs are almost limitless. And with a country like India, it could absorb all the capital that we or anybody else is prepared to send there, particularly if we would bring some managerial skills along with it. The populations are so vast that I think the needs are really unlimited, and it's hard to set up absolute priorities. In each case we have to work out the best mix that we can get for the money that we can afford to invest in any one of these countries.

QUESTION. I was wondering if you could comment on the role or function of labor unions, in the industrial development of the United States, and relate this process to the development of some of the underdeveloped countries.

KASSALOW. Let me talk first of the developing countries. One of the socio-political needs of the developing countries, particularly because such states are going to be so tremendously powerful, is to build so-called intermediate associations, or to preserve them, so as to diffuse power in the society. There have to be channels through which people can participate. These kinds of associations become very significant and very important if there is to be democratic development rather than purely authoritarian or totalitarian development, which is the normal tendency of these states.

In this sense the trade union, along with other types of associations, can be a vital and a fairly important institution in the nation. It has other roles to perform as well, in this process. The process of development is going to be a grueling and difficult one; rapid urbanization and rapid technological change cannot be an easy, pleasant process. It seems to me the trade union has a kind of a lightning rod function to perform in this process by channeling discontent and taking the heat out of some of the industrial tensions which must inevitably develop in the course of this speeded-up process.

But we also begin to see a glimmer of some new functions in Asia and Africa for trade unions. If they can function in this intermediate association capacity, perhaps they can take on part of the problem of urbanizing, softening some of the urban impact upon their members who will be coming from all over the countryside or from villages, in many instances. Maybe they can become part of the instrumentality whereby houses are built for workers, or vocational education is extended to them. These are functions for trade unionism which did not have much currency in Western industrialization, but perhaps the process is going to be so different here that they can be used as institutions in this respect also.

But let me make one most important point. Under the influence of world opinion, channeled through the International Labour Organisation (ILO), there are already or will be soon almost everywhere in these countries something called trade unions. Whether it's in Asia, Africa, or Latin America, there are trade unions which are either demanding, fighting institutions, or trade unions which are arms of the government to help mobilize the workers or help bring them into the process of national development or economic development. The interesting thing in Africa, for example, is that there were trade unions in nearly all of the African nations after World War II, and a number of them played an important role in the national liberation struggle, associated with the political parties. As the countries gain independence, the new political leaders are fearful of these trade unions. They look upon them as a potential opposition force, and small as they are, they have helped to overthrow several governments in Africa already.

Now, what do they do? You know, in no country in Africa that I can think of where there have been trade unions has the government eliminated them. What it has done is to try to take them over, incorporate them into the state, or build them up, because given the ILO, the prestige it has in the world, given the fact that modernism seems also to encompass trade unionism, every state almost has to have its labor front or its labor organization. So we're getting a new phenomenon, in terms of trade unionism, in many of these new societies. In the older countries like India and Japan, where there are more traditional class structures, the trade unions have had to some extent the role of combat organizations they've had in Western development, and in the case of Japan, revolutionary Socialist organizations as well. But we are going to see them in the new nations on a scale far greater than they were known at a comparable stage in Western development—I think we can count on this. The real problem is, can they work out constructive, useful functions, while maintaining a certain amount of their own viability and independence?

Alternatives to Technology

LOREN C. EISELEY

ONE OF THE THINGS that sometimes haunts me on occasions such as this is Emerson's remark, made in the dawning period of modern science, about the terrible questions men are beginning to ask. I have thought upon the depth of his perception and its timeless quality, because each year the questions that beset us grow more imperative and difficult to answer. As Sir Isaiah Berlin has already warned us, the number of variables which entangle themselves with human history multiply in a kind of geometric progression that challenges all simplistic explanations.

Even at these meetings we find ourselves before a variety of conflicting crossroad signs. We are to return to Fortress America and conserve our resources; we are to move out of Fortress America; we are to save the world; we are to feed everyone. And if we do not, we will be very guilty men. I have a feeling that perhaps we are failing to take account of Dr. Berlin's caution, and that it might be advantageous to explore man's evolutionary history in the attempt to see just where we stood.

We have heard a great deal in the last few decades about "one world." Here is where our modern ideology enters in—we are always assuming that it is *our* world which is expanding. I will not argue the fact that our technological culture is diffusing at a rapid pace. I would like to ask, however, a few questions

LOREN C. EISELEY IS PROFESSOR OF ANTHROPOLOGY AT THE UNIVERSITY OF PENNSYLVANIA.

about this world. In the course of two million years, something unique has appeared.

I remember the first time that I ever saw an archeological stratum exposed in a mud bank. There is pain in the contemplation of not only the problems of today; there is pain in the contemplation of the problems of the past. No society in history has ever before been so conscious of the failure of other civilizations, and no civilization in history has ever before ranged farther into the past, or sought more intensely to pierce the future. It is indeed a painful process. Even as a young man I was touched with sadness as I gazed upon the crushed, calcined debris that represented centuries of effort on the part of a forgotten people. In that stratum was all that remained of human thinking and human effort at a given place and time. A little farther away was another stratum, another time horizon where another set of men had striven, as we are striving now. We can go all the way back through time, far below the emergence of man. The paleontologists will tell us that the majority of past life on earth has perished, that what is represented around us is perhaps less than 10 percent of earth's once living forms, most of which have died away or been transformed into something else, just as man has been transformed.

In terms of human evolution, there are two ways of looking at the process, and both of them, to a degree, are represented among scientists today. There is one view which is inclined to say that the only way in which life alters or changes is when change is forced upon it. There is a certain mechanical aspect to this, an assumption that man did not become what he was until certain changes in his environment had occurred. That the proposition has a degree of validity there can be no doubt.

But there is another point of view, which is that in life itself there is a centrifugal dynamism of sorts, not just in man but in all living creatures. It does not wait upon its environment, instead it intrudes farther and farther into it, experimenting on its own. I would like to tell a little story in this connection. Some friends of ours, who had a house out on the edge of town, went away for a two-weeks' vacation. When they returned they found something very interesting had occurred. They found they had a visitor from another world. The visitor was fortunate in the sense that my friends had by chance left a little water dripping from a faucet in the kitchen. They had also left accessible a few ends of bread and other food. Somehow or other, making a descent through space down the chimney, had come a flying squirrel who, as it turned out, had decided to spend the winter with them. Life experimenting, you see.

I could tell endless stories of this kind about what animals do. I have encountered squirrels in subways and snakes climbing through second-story windows along drain pipes, and things of this sort. In other words, life is always pushing outward against its environment, experimenting on its own, not necessarily waiting for something to force it in that direction. I suspect this inner drive culminates in the nature of man.

At any rate, the squirrel proceeded to set up housekeeping in a series of drawers in the kitchen cabinet. The first drawer he used for sleeping, tucking himself away under a napkin in the daytime. In the second drawer—this sounds incredible, but it is true—he began, in a capitalistic fashion, to accumulate wealth, and it was not all just food. He was a curious and interested squirrel. He scurried around collecting some old forks and spoons and odd fragments of boxes, things that had no direct food appeal to him but that for some reason or other he found attractive, and these he stored carefully away, as a little treasure trove, in the second drawer.

The family arrived home just in time to help the fellow. Much to their astonishment, they saw a little head peering over the back of the sink, with big brown attractive eyes, and thus they discovered the squirrel. They treated him politely (they were kind people) and he reciprocated. They helped out with bits of food and they maintained the quiet necessary for his day-time sleep. In the spring the squirrel scampered back to his world through the open door.

This is an example of life experimenting. It has its appealing side. We, I would like to remind my audience, have some very curious aspects to our anatomy. We have hands. I think we sometimes forget that we got them through a series of long and curious travels through a forest attic. The hand is an old general-ized structure intended originally for climbing. One aspect of that forest attic—prologue to the development of human culture —is the fact that a creature traveling in trees is confronted with a waving laboratory of tools. He cannot travel along in an auto-matic fashion. He has to watch things which are moving. He has to have keen stereoscopic vision. He has to seize and hold mov-ing things. And so, in a rough way, we might say that these creatures of the arboreal attic are tool-manipulators before they realize at all that they are manipulators of tools.

Written along the fissure of Rolando in the motor center of our brains is an exaggerated representation of the thumb and forefinger. The center for the control of the tongue exists also in the brain. It is interesting to study the neurological charts be-cause so much of man's innovative curiosity involves the tongue. Also in that little box in which the individual brain is carried there are provisions for high auditory discrimination. Within the skull box is located the specialized control centers for the hand, the forecasting eye, and that ear which is attuned to sound. I would like to remind you that one of the peculiarities of man is that his first great source of energy is a puff of air, the word. The forecasting eye and brain which found its way through the movements of the forest branches has about it an anticipation of the future which, once you have begun to chop up and catego-rize your world and to name it, brings into clearer vision the past, the present, and the future.

So, as the anthropologist would say, first of all we were

brachiators, but we got down out of the trees before we were too specialized in that direction. We made some other remarkable adaptations, such as upright posture. But is it not true, to a degree, that with the forecasting eye and communicative use of the tongue, that we now brachiate among ideas which also waver, change, and are equally as difficult to grasp as the waving branches in the forest?

Scarcely had we arrived in the earliest phase of what we might call human development before we began to substitute for the lost world of pure instinctive behavior a second nature to which we could become conditioned, and establish a new kind of order in our lives, even as we escaped the old world of nature. That new order is the world of society and its institutions. We also began to acquire the ability to abstract—be apperceptive, as the psychologists would say. In other words, we could hold ideas, symbolically, and have a conception of ourselves, as distinct from nature. It is possible today for the speaker to see that he is struggling with language, worried about a presentation, even as he proceeds with his discourse. There is an "I" which recognizes an "I," and thus transcends itself.

This new world, this world of culture which man entered, is involved with institutions, and institutions, though they have been defined in many ways by the social scientists, are essentially those aspects of the social structure which (like the corporation) endure beyond the lifetime of the individual. They are common forms of order. They are, it has been well said, the bones of the societal body. We exist, most of our lives, within the interstices of institutions, whether economic, religious, political, or familial. We have had nothing to do with their formation, even though we know that these institutions can change, and in some eras of human history have changed more rapidly than in others. We have emotional, deep-rooted attachments to these institutions. Much of our activity within them sinks into the subconscious. To a surprising degree they order our lives.

Not long ago a friend of my wife suffered a small stroke while she was in church. She had been engaged in some vital religious meditations, when she experienced an attack of amnesia. She retained no memory of what she did, but somehow or other she did everything that she was supposed appropriately to do under these circumstances and went home. Only later did she realize who she was. In other words, so deep was the ingrained behavior of the woman's normal routine activity that all of it was carried out in an automatic subconscious fashion. In some such way many of our routine institutionalized acts tend to sink below the conscious level.

Two million years ago, we begin to have evidence of a creature that we can doubtfully call human, bipedal, smaller than ourselves, and already capable of shaping simple tools. We observe marked distinctions among these creatures, variations more distinct by far than persist in the modern races. As to which of these forms survived to become the direct ancestor of *Homo sapiens*, at present we can only speculate.

We finally reach a point where our own species emerges. *Homo sapiens* is a specific name which covers all the living races of whatever color. Somewhere along this ancestral line, either by interlinkage, interbreeding, or other means, the physical differentiation of man began to attenuate toward a common specific point. As man entered, or, we might better say, created this strange institutional world of culture, one of the most momentous steps that a living creature has ever achieved on the planet, his brain, had become a dissolver of substance. The brain could do so by manipulating symbols. It could take rock and turn the rock back on the nature from which it was derived. It could extract ores, manufacture metals, seize upon all manner of material substances, and literally melt and mold them into other shapes.

Long before the dawn of civilization these potentialities were present in that human brain. The woman who rocked an Ice Age child to sleep in a cavern, the man who painted on the walls of that cavern the beautiful Paleolithic art which we now admire, would be capable, if they were sitting here among us, of all that we can achieve. In other words, the latent abilities in that brain were enormous. What it may further achieve by cumulative effort we do not know.

But here is the paradox. The human form, originally so variable, has been narrowed into one new channel whose primary emphasis involves the rise and development of the human cortex. An enormous spectrum of possible behavior, of possible adaptation, a spectrum so broad that it clashes and contradicts, is the product of that brain. Each society, by conditioning of the young, by the nature of its beliefs which have an evolution of their own, has selected a certain segment of human behavior for exploitation or emphasis. So there has arisen a variety of insubstantial roads into the future, even though physically men are much more alike than they had been in the beginning. One of the things that troubles us now is that in this great whirlpool of modern civilization, this centrifugal power, represented by modern science and technology, Western man is sweeping away all other societies. The process is still comparatively close to its origins.

The basic problem, as one sees it here, involves something strangely analogous to mutation in the biological realm. Let me explain why. Sex, you know, was a marvelous invention. It promotes diversity and emergent novelty among men, just as it does in the animal world. It is responsible for that whole aspect of life which we speak of as evolution, whether in plants, animals, or men. Involved in it, also, are the temperamental qualities of people, over and beyond the way that these may be influenced by social conditioning. One of the things which mutation does, so far as the lower animals are concerned, is to open doorways. It enables life to penetrate new corridors of existence. Changes take place. This dynamism of life goes on, this drive toward expansion, whether it involves parachuting down chimneys, climbing trees or climbing down from trees, or entering, as man long ago entered, the grassland corridor which was destined to direct his footsteps in so strange a fashion.

After one of these unique biological inventions has been made, what we call adaptive radiation takes place. Based upon an original biological invention, a multiplicity of diverging forms spread out. In the social realm, one might observe that something of a comparable nature took place in the emergence of societies adapted to particular ecological circumstances around the world. These ways of life have long histories, and they have continued into the present. Now, in the age of the atom, in the age of this great expanding whirlpool of technology, there still exist marginal remnants, remnants from the Stone Age, people who look at us from a remote distance in time even though they are really existing on the same time horizon as ourselves.

As Americans we are frequently impelled by the feeling that our own society is superior, even though we have gotten over some of the naive ethnocentricity of the earlier voyages, first discovering the width and the diversity of the world and of human cultures. We feel that no one can possibly be happy unless he has everything that we have, behaves in the same fashion, and believes in democracy and votes.

It is said that the anthropologist is a romanticist. He goes out to some primitive people. He likes them. He learns about them, he lives among them. Then he comes home and underneath, psychologically, he wants to return to the Stone Age. Let us leave that aside, because something more than romanticism is involved here. Life has evolved by mutations. The moment we concentrate life within a single animal form, we are placing our hopes for life, our hopes for its successful continuation, all upon a single species. Nature rarely, if ever, does this. Even in intricate cases when some things are elaborated and specialized, nature always manages to keep around a few generalized animals that may turn out to be the real masters of the next epoch. It happened in the case of the dinosaurs. Their relatives, the smaller therapsid reptiles, gave rise to the true mammals who, in turn, engaged in a highly diverse radiation into the life zones abandoned by the mysterious disappearance of the giant dinosaurs.

Similarly, in human societies, if we were to examine them throughout centuries of time, we would find the ways of life, the value systems, quite different. The cultural "mutations," one might say, would have emphasized a certain segment of human behavior, exploiting it in a particular fashion. Not "one world" would be the catchword, but a variety of worlds.

I have sat, at times, among people of other cultures, now aware of our own, and perhaps unable to say what is coming. They have said to me—some of them highly intelligent people in their own way—"We do not want your transistors. We do not want your system of government. We do not want your religion. We do not want what you have precipitated. We do not like your games." But still, the erosion goes on. The whirlpool reaches farther and farther and farther, as the communications network reaches out. Whether or not it is wise or augurs well for men in these other societies, their way of life is doomed.

There used to be more than one way into the future—in evolution, through mutations, and in a sense, in terms of the solitary civilizations which have flourished and existed in the past. But now these diverse worlds are increasingly alike even when they conflict. They are in the same rotating whirlpool—that of Western technology. An event is taking place which would never occur under biological circumstances: a single monstrous "animal" is being born on which we are placing all our hopes for mankind. With all the ingenuity that we still possess, would it not be well for us to ask, before the gateway closes forever: are we sure where we are going? Are we sure that what we are doing is the right thing to do? Are we even sure that it is our version of science and technology that will survive? There was another version not long ago, that of Nazi Germany.

There is another thing that happens in societies, not alone in our own and not alone in the case of the single monstrous "animal" I have described, but in any division, any part of this spectrum of activity, this assemblage of institutional forms. This is what the social scientist describes as involution—a kind of exaggerated growth constructed around a mythos to which the rest of the society becomes a mere adjunct. Once that has happened, it is very difficult to divert societal energy elsewhere. Some of these involutions are different from others and have different psychological effects upon their people. One can have extremely militaristic and aggressive societies. In that case we like to say, "This is human nature." One can also have balanced, peaceful, centripetally oriented societies which do not have this drive outward. As wealth accumulates in civilizations, it can be expended in many different ways. It can be expended upon a religious system, so that excess wealth and energy is drained in that direction, as in the giant sepulchers of ancient Egypt. Or it may be expended upon a great military machine. Or a caste structure may develop in which social boundaries become watertight, constricted, and the society static. There are many ways in which these curious involutions can occur.

Our own culture possesses the same potentiality. Once the energy and the thought of a society begin to drive in a particular direction, it becomes ever more difficult for any man singly to divert the process or to call a halt. The whole of Western societal impetus now is towards the creation of an enormous industrial–military establishment, with all that that involves. More and more of our wealth is poured into it. It employs, directly or indirectly, a larger and larger body of people. And so the monstrous "animal" begins to take on specific shape, even though at present the form may not be totally discernible.

The world is old, how old we are just beginning to realize, and throughout the long course of both geological and biological time the world has slowly altered. Continents have moved up and down. The intrusion of epeiric seas upon the platform of the continents has come and gone. It has been a slow play of chance, of contingency, as the biologist phrases it. The change has proceeded slowly enough that life, through these endless millions of years, has been able to make its readjustments. With the introduction of the reshuffling of the genes in the sexual system, however, a greater degree of variability was introduced. Those who play cards will better understand this aspect of probability. Contingency itself is evolving at a quickening pace. The vibrations it induces are increasing. When the cultural level is reached, the level of social as distinguished from genetic heredity, the pace steps up, begins to move faster and faster. At first, in cultural islands of isolation, these developments are hidden, evolving alone to a considerable degree.

Some three centuries ago science, in the sense we know it, began to be taken seriously. And in just three centuries, the amount of chance, contingency, the play of innumerable forces, has mounted in a kind of geometric progression, so that there are many who will remember things that have already vanished.

Institutions have histories or they are not institutions. As I have said, they are the bones of the societal body. When they dissolve, we no longer have a viable society. Over the generations, institutions slowly change but they have to have a certain amount of stability to sustain societal order. Flickering and dancing in the play of our great machines, however, and haunting our scientific establishment, is an unseen invisible elf to whose whims we are subjected in an ever-intensifying fashion—an elf we call *change* or *contingency*. I am sure that at the moment of my death, except for a certain linguistic continuity, I will look my last upon a society so different that I will feel as though misty centuries have passed since my birth. The question is, as the contingency factor builds up in this fashion, how long man can adapt to the rapidity of social change, and how well the single great society—or as I prefer to call it, the monstrous "animal"—can sustain itself and at the same time sustain our human freedoms.

When all the world is part of the great society, from East to West; when these uncounted millions of human beings each demand their two cars or their three cars and the many other things which our society enjoys; when all nations have factories, and all are turning out these goods—how many resources on this increasingly impoverished planet earth will it take to satisfy this material demand? I merely ask. I do not have an answer.

There was a magician, once, who, when he was approached by a young apprentice for advice in connection with the raising of demons, gave a remarkable answer. He said, "Do not call up that which you cannot put down. Have always the words for laying close at hand." I am afraid that we have forgotten the words for laying. They should, I think, be close at hand, somewhere.

In the time of the great power blackout in New York a strangely symbolic episode took place. There was a man on one of the high floors in a darkened building in Manhattan who set out to investigate with a candle. As he progressed along the dark and drafty corridor, with his upheld candle, he did not know that he was in some fashion walking with his little flickering light of knowledge into the windy spaces of the universe. He proceeded with his candle until he came to a door. Because he was a modern man, not skilled in the technology of the forest, not afraid of demons, not warned against traveling unknown paths at midnight, he opened the door boldly and walked in. They found him, three days later, with the candle upon which he had relied still clutched in his hand, at the bottom of an elevator shaft.

The candle had proved inadequate in the end. Man, as he builds this new enormous civilization (and I am enough of a realist to know that he will not cease to build it) is intensifying the area of contingency impinging upon him—good, bad, indifferent; and the candle increasingly wavers in his hands.

We are within the structure. We will make future-oriented choices, good or bad, that will persistently affect us, over and over, until the speeded pace of change can be heard knocking in our exhausted hearts. Emerson did more than prophetically speak of terrible questions to be asked. He said even more conclusively, "The spirit builds itself a house, but afterwards the house confines the spirit. Beware of the kind of house you build."

OPEN DISCUSSION

QUESTION. I have mingled professionally with sociologists and other social scientists, and I find you often hear them say— they're very candid, very honest—that sociology today is at the level that physical science was in about 1620; it is just beginning. I suggest that what is needed is for sociologists to push themselves very hard, so as to redress the imbalance, to bring sociology up to the level of the other sciences, so that human beings can learn a little better behavior. Do you agree?

EISELEY. I would say we need a little caution about this business that sociology has made no advances since 1620, or that we cannot predict or project in the way that the physical scientists can project. And let me say, I do not believe we ever will. I think this is another kind of science. Though I'm not a sociologist, I think anthropology overlaps sufficiently into that area so that I shall take the opportunity to comment on your proposition.

I think we worship predictability too much. Men in the hard sciences frequently speak as though predictability, sureness of prediction, were the primary object of science. It may be, under certain circumstances. But I think we should remember we are dealing with a creature who, as remarked before, transcends himself, looks at himself, has in a sense a mirror in him, and changes even as we look at him, because of ideas and styles and fashions. Therefore, trying to lay down some precise physical–social laws about him is a very dangerous procedure.

He partakes, in other words, of that same intangible quality of contingency that we've been talking about. There play upon him such a vast variety of influences, calculable and incalculable, that I do not think that the course of civilizations can be laid out in this fashion. They are too unpredictable. Because I see no way in which they could be made predictable, I wouldn't urge for a single moment the expenditure of vast sums of money on that type of research. Maybe I'm just being old-fashioned here, and I will leave it to others more experienced in this field to comment if they wish. But I think that this kind of science—and I still call it a science, if for no more reason than that it opens the doorway to consideration of man, not in terms of one society, but in terms of many—is worth pursuit.

QUESTION. Presuming atomic destruction of all humankind, except for vegetation and things that live in the sea, how long do you think it would take, anthropologically, to build back to the stage that we now have?

EISELEY. We probably never would. If we assume that man is totally eliminated, we must remember that he is a product of biological contingency as he now stands. He does not appear everywhere automatically. We could spend a half hour or so on that subject, but I would merely say that the probability is very low for the emergence of the same line of evolutionary development twice, because of the total contingencies specifically involved in the production of man. This would mean that whatever forms came after such a destruction and whatever direction they took, they would not be man.

In Australia, which is dominated primarily by marsupial fauna, which are only very remotely related to ourselves, there has been no emergence of anything comparable. Also, in spite of certain kinds of parallelism in nature, you cannot name one single creature that has again appeared on this planet, once it has disappeared. Now bear in mind, I did not say that in terms of certain kinds of adaptations, you may not get an empty corridor filled by a new form of life. You can get warm-blooded mammals returning to the sea, as in the case of whales. But the adaptations they make will not be fish adaptations. A whale breathes air. He's wrapped up in warm winter underwear in the shape of blubber to protect his warm blood. He has developed a whole series of new adaptations for the water environment, even though he has come off the land and returned to the ancient birth environment from which we emerged. This, I think, is a pretty good demonstration of the general principle of the irreversibility of evolution.

A POSTSCRIPT BY PROFESSOR EISELEY

During the discussion immediately after Sir Isaiah Berlin's paper, Professor Eiseley made a comment on the significance of social change that seems even more appropriate as the conclusion to this volume than it did as a high point in that discussion:

EISELEY. I thought I detected among us today a feeling that ideology was necessary to change, and that change in all aspects was somehow necessary. Here again, I think, we're speaking (and I don't want to exaggerate this) from the standpoint of a Western society habituated in almost all aspects to change, whether that change is significant or necessary, or valuable to us, or is in fact, quite the reverse.

I can't help but remember some years ago spending a period of time in a remote village in another country. Past the village ran an asphalt road installed by that country, probably with our assistance. It was mostly an unused road, and I noticed that in the evenings people came out and sat by the road enjoying the warmth (this was high mountain country) radiating from it.

They were quiet people. They were content. No one had come to tell them about their deprived state, yet. I realized that their death rate, especially their infantile death rate, was high. I realized that if I had lived my whole life in that society, I would probably have been dead by that time. But for all that, there was a quiet among these people, an acceptance of life, that I, a restless, febrile sort of individual coming from our highly urban, complex society, found most comforting. And I would like to comment that before we settle on this endless course of change in all things, let us be sure what change we want; whether change in all aspects is necessary. I know we live in a highly dynamic culture that is not going to alter because of anything I say here. On the other hand, it seems to me that through the ages some of the great thinking personalities have been able to talk across the gaps between cultures and between centuries simply because they have passed beyond the ephemeral qualities of change into some other kind of domain that involves what is enduring in humanity.

THE ENVIRONMENT OF CHANGE:

CONFERENCE AT ONCHIOTA CONFERENCE CENTER,
STERLING FOREST, NEW YORK,
JANUARY 12–15, 1966

Participants from the Business Community

Edward B. Bates
Executive Vice-President
The Connecticut Mutual Life
  Insurance Company

Lyle L. Blahna
Vice-President: Marketing,
Media, and Research
MacManus, John, and Adams,
  Inc.

Robert A. Buck
Vice-President: Marketing
Pet Milk Company

Stanford Calderwood
Vice-President
Polaroid Corporation

George Couch
Vice-President: Marketing
  Operations
Anheuser-Busch, Inc.

Joseph R. Daly
Senior Vice-President
Doyle Dane Bernbach, Inc.

Arthur E. Earley
Vice-President: Creative Services
Meldrum and Fewsmith, Inc.

John L. Fleming
Vice-President: Public Relations
  and Advertising
Aluminum Company of America

E. Edgar Fogle
Vice-President: Marketing
Union Carbide Corporation

F. William Free
President
The Marschalk Company, Inc.

Nelson B. Fry
Executive Vice-President
Allegheny Airlines, Inc.

C. Gus Grant
Group Vice-President
Ampex Corporation

J. Richard Grieb
Director: Sales Development
Maxwell House Division
General Foods Corporation

Richard K. Jewett
Director of Advertising
Pitney-Bowes, Inc.

William G. Kay Jr.
Vice-President: Frozen Food
  Division

Pepperidge Farm, Inc.

John P. Kennedy
Vice-President: Director of
     Marketing
Bristol-Myers Products Division
Bristol-Myers Company

Walter J. McNerney
President
Blue Cross Association
Chicago, Illinois

Robert A. Sandberg

Director: Advertising and Public
     Affairs
Kaiser Aluminum and Chemical
     Corporation

Irving Scharf
Director of Marketing
Seagram Distillers Company

Emil Seerup
Vice-President: Air Services
     Department
REA Express

*Speakers and Members of Columbia University Seminar on
Technology and Social Change*

Sir Isaiah Berlin *
Oxford University
Oxford, England

Thomas E. Cooney Jr.
Program Associate
The Ford Foundation

Loren C. Eiseley *
Professor of Anthropology
University of Pennsylvania

Eli Ginzberg *
Hepburn Professor of
     Economics
Columbia University

Walter Goldstein
Professor of Political Science
Brooklyn College

Norman Kaplan
Program of Policy Studies in
     Science and Technology
George Washington University
Washington, D.C.

  * Speakers

Everett M. Kassalow *
Professor of Economics
University of Wisconsin

Melvin Kranzberg
Professor of History
Case Institute of Technology

Dean Morse
Assistant Professor of Economics
Columbia University

Mario Salvadori
Professor of Civil Engineering
     and Architecture
Columbia University

David Sidorsky
Associate Professor of
     Philosophy
Columbia University

Frank Tannenbaum
Professor of History
Columbia University

Aaron W. Warner **
Joseph L. Buttenwieser
Professor of Human Relations
Columbia University

I. I. Rabi *
University Professor Emeritus
Columbia University

Participants from Time, Inc.

Bernhard M. Auer
Senior Vice-President

Robert C. Barr
Advertising Sales Manager

Herbert E. Brown
Manager, Creative Contact
Advertising Sales

James W. Cobbs
Marketing Services Director

Richard E. Coffey
Promotion Director

Sheldon Cotler
Promotion Graphics Director

Walter Daran
Photographer

John J. Frey
Assistant Publisher

Robert S. Fuiks
Manager, Corporate and
  Industrial Advertising

Robert C. Gordon
Advertising Sales Director

John A. Higgons III
Manager: Consumer Goods
  Advertising

Dean Hill Jr.
Regional Advertising Sales
  Manager

John P. Keller
Public Affairs Department

Lawrence E. Laybourne
Assistant Publisher

Marshall R. Loeb
Senior Editor

John A. Meyers
New York Advertising Sales
  Manager

Kelso F. Sutton
Business Manager

Robert D. Sweeney
Public Affairs Director

James A. Thomason
General Manager

George W. McClellan
Advertising Sales Manager
Pittsburgh, Pennsylvania

Albert G. Watkins
Advertising Sales Manager
St. Louis, Missouri

* Speaker

** Chairman

THE ENVIRONMENT OF CHANGE:

CONFERENCE AT AIRLIE HOUSE, WARRENTON,
VIRGINIA, SEPTEMBER 18–21, 1966

Participants

Bernhard M. Auer
Senior Vice-President
Time, Inc.

William J. Bailey
Vice-President and Assistant to
the President
Carrier Corporation

H. O. Baker
Sales Manager, U.S.A.
BOAC

Robert C. Barr
Advertising Sales Manager
Time

James A. G. Beales
Director of Marketing
Calgon Corporation

John Bowles
Vice-President: Marketing
Beckman Instruments, Inc.

Dr. Jacob Bronowski *
Senior Fellow and Trustee
The Salk Institute for Biological
Studies

James W. Cobbs
Director: Marketing Services
Time

Robert P. Fisler
Director: Promotion
Department
Life

J. Wendell Forbes
Special Projects Manager
Time-Life Books

John J. Frey
Assistant Publisher
Time

Robert S. Fuiks
Manager: Corporate and
Industrial Advertising
Time

Eli Ginzberg *
Hepburn Professor of
Economics
Columbia University

Charles L. Gleason Jr.
Personnel Director
Time, Inc.

Robert C. Gordon
Advertising Sales Director
Time

John H. Goy
Vice-President and Manager
San Francisco Main Office
Bank of America

Dean Hill Jr.
Advertising Regional Manager
Time

Jack K. Howard
Division Manager
Technology Transfer Division
 (TETRAD).
United States Rubber Company

Henry H. Hunter
Vice-President: Marketing
 Services
Olin Mathieson Chemical
 Corporation

J. Emmet Judge
Vice-President: Marketing
 Services
Westinghouse Electric
 Corporation

Everett M. Kassalow *
Professor of Economics
University of Wisconsin

John P. Keller
Public Affairs Department
Time

James C. Keogh
Assistant Managing Editor
Time

Frank J. Lionette
Vice-President: Advertising
Howard Johnson Company

Marshall Loeb
Senior Editor
Time

James B. McIntosh
Senior Vice-President
New England Mutual Life
 Insurance Company

John A. Meyers
New York Advertising Manager
Time

Dean Morse
Assistant Professor of Economics
Columbia University

Keith Olson
Marketing Services
Time

I. I. Rabi *
University Professor
Columbia University

William H. Roberts
President
York Division of Borg-Warner
 Corporation

Edward Russell Jr.
Vice-President: Marketing
 Services
Champion Papers

M. P. Ryan
Advertising Director
Allied Chemical Corporation

Mario Salvadori *
Professor of Civil Engineering
 and Architecture
Columbia University

Robert M. Schaeberle
Executive Vice-President
National Biscuit Company

David Sidorsky
Associate Professor of
 Philosophy
Columbia University

Frank Tannenbaum
Professor of History
Columbia University

John C. Thomas Jr.
Los Angeles Advertising
   Manager
Time

James A. Thomason
General Manager
Time

Turrell Uleman
Director: Commercial Research
   and Development
Pittsburgh Plate Glass Company

Aaron W. Warner **
Joseph L. Buttenwieser
Professor of Human Relations
Columbia University

Walter Daran
Photographer
Time, Inc.

Ellen Stancs
Public Affairs Department
Time

* Speaker   ** Chairman